PREFLIGHT PLANNING

PREFLIGHT

MACMILLAN PUBLISHING COMPANY
New York

COLLIER MACMILLAN PUBLISHERS
London

PLANNING

RON FOWLER

Illustrations by Jan Avis
Photographs by John Tate

Copyright © 1983 by Ron Fowler

All rights reserved. No part of this book may be reproduced or transmitted in any form or by any means, electronic or mechanical, including photocopying, recording or by any information storage and retrieval system, without permission in writing from the Publisher.

Macmillan Publishing Company
866 Third Avenue, New York, N.Y. 10022
Collier Macmillan Canada, Inc.

Library of Congress Cataloging in Publication Data

Fowler, Ron.
   Preflight planning.

   Includes index.
   1. Aeronautics—Safety measures. 2. Navigation (Aeronautics) 3. Airplanes—Piloting. I. Title.
TL553.5.F68   1983   629.132'52   83-729
ISBN 0-02-540300-1

---

10 9 8 7 6 5 4 3 2 1

Printed in the United States of America

TO
Lindsay

Contents

Acknowledgments ix
Introduction: Preflight Planning and Professionalism 1

THE PILOT

CHAPTER 1. *Establishing Personal Weather Minimums* 9
CHAPTER 2. *Gauging Your Accident Potential* 53
CHAPTER 3. *Complying with FARs* 66
CHAPTER 4. *Keeping the Pilot Flight-Ready* 74

THE AIRPLANE

CHAPTER 5. *Evaluating Aircraft Performance* 113
CHAPTER 6. *Checking the Plane for Flight* 134
CHAPTER 7. *Complying with FARs* 161
CHAPTER 8. *Qualifying the Plane for IFR Flight* 165

THE ENVIRONMENT

CHAPTER 9. *Relating the Weather Briefing to the Flight* 177
CHAPTER 10. *Preflighting the Destination Airport* 194
CHAPTER 11. *Compiling the Radio Frequency Log* 203

CHAPTER 12.	Preplanning the Navigation	220
CHAPTER 13.	Complying with FARs	232
CHAPTER 14.	Evaluating the Environment for IFR Operations	235

Conclusion: A Pilot's Responsibility   248

Index   251

Acknowledgments

Many preflight suggestions in this book reflect ideas of the instructors and pilots I've worked with for almost twenty years. Among that group, I am indebted to pilots Jim Dubick, Penny Wilson, Jim Brady, Beverly Morton, Bob McLain, Jim Connell, Ed Banks, John Tate, Ed Karvonen, Dick Wilcox, Tommy Smith, George Jamieson, Bill Smith, Dan Russell, Ed Gomis, Bill Sargent, Bryant Bouslog, Ed Joehrendt, Ted Huckabee, Billy Lee, Bud Sypher, Fred Tilden, Frank Wignals, Ken Murray, Mary Blackwell, Miles Trylovitch, Roy Beatty, Mac Barksdale, and Will Shaw.

I am also appreciative to those pilots whose misadventures have turned up in this book as the "horrible examples." I hope they don't mind too much. To even the score, I've put in an equal number of my own foolish mistakes.

The brotherhood of pilots, some say. Others call it the fraternity of airmen. I'm not sure which is correct, but this I do know: To a pilot, the friends who matter most have to do with flying. My sincere thanks to all.

Special thanks also go to Triangle Reprographics for helping with the graphics; to Helen, my wife, for typing, retyping, and re-retyping those pads of unreadable handwriting; and to daughter Betsy for editing the ramblings of a pilot-turned-writer into a form more acceptable to a publisher.

PREFLIGHT PLANNING

INTRODUCTION: **Preflight Planning and Professionalism**

THE FEDERAL AVIATION ADMINISTRATION'S *Airman's Information Manual* lists by frequency of occurrence the ten most common causes for pilot error and accidents. The first cause is listed as "inadequate preflight preparation and/or planning."

Much of a pilot's competence and much of a flight's success depend upon adequate preflight planning. Both competence and success underlie the basic truth of flying.

The basic truth of flying is simply this: Plane and sky grant no special consideration. Whether we walk the earth as good guys or bad, poor or rich, wise or foolish makes no difference. When we enter the cockpit we leave behind our earthbound differences. We touch the controls as one—a pilot. If we are competent, our flying will probably succeed. And if we are incompetent, our flying will probably *not* succeed.

A pilot builds competence in three areas of expended effort: skill, judgment, and knowledge. Of these three, knowledge is the cornerstone, for without knowledge our skill is easily misdirected, and our judgment has nothing to draw upon.

Our knowledge consists of two intermingling pools. First, there is our reservoir of aeronautical knowledge; the background weather, aerodynamics, and the hundreds of other facts of flight that are in authority every time we leave earth behind. The second area of knowledge consists of those changeable facts that we must gather before each flight. And it is in this area—preflight planning—that we often let ourselves down.

Many pilots feel slightly uneasy when flying cross-country. In most cases they are not as concerned about any lack of skill on their part as they are about the prospects of suddenly meeting the unexpected, of being "caught by surprise." And most of these pilots never relate their lack of in-flight confidence to inadequate preflight planning.

Unless we take positive steps to prevent it, our own "preflight" is apt to consist only of a weather briefing and a light once-over inspection of the aircraft. Then we stand a good chance of the uncomfortable (and sometimes hazardous) moments when we are suddenly faced with unanticipated situations during the flight. A pilot who suddenly needs to scramble through a jumble of charts to find a needed frequency, for example, forgets to fly the airplane or scan for traffic. It seems a simple matter in the reading, doesn't it? But if the need for the elusive frequency is pressing, he is liable to become frustrated in his search—and his antics scare the pants off his passengers.

The unaware pilot who quickly needs an unanticipated alternate plan gives us another example of inadequate and unprofessional preflight planning in action. Picture yourself in the passenger seat as the pilot nears the completion of an afternoon's cross-country flight into unfamiliar territory. An unexpected head wind has drawn the fuel gauges down to the quarter mark and pushed the destination another thirty minutes ahead. A setting sun and dwindling dew-point spread have steadily cut visibility to five miles and given dusk a head start. Fast moving scud has forced the plane to a thousand feet over the terrain, whose features are already half hidden in the gathering gloom. And a scant five miles ahead, the pilot sees a blue-gray curtain of precipitation that cuts off a direct flight to his destination. Moments are tense as the unprepared pilot tries to put together,

A familiar sign. Many pilots have heard the term "total awareness," but few have experienced it.

from scratch, a change of action as critical decisions face him. And the propeller keeps pulling him ahead at 150 knots toward potential pilot error.

But pilots do have a choice. Every pilot can remove the "unanticipated" from his flying, through total preflight planning. Total preflight planning means knowing all of the significant factors that can affect any flight (short or long), researching and evaluating those factors as they pertain to each flight, and making plans to meet them with a series of alternate actions that may be needed.

There is a parallel (and often misunderstood) ideal that goes hand in hand with "total preflight planning"—"total awareness." Simply put, total awareness means never being caught by surprise in the cockpit. It is a concept that many pilots hear about but few experience. The majority of the pilots who never attain total awareness let it elude them simply because they do not understand its simplicity. They feel that it is "theory stuff." Yet it is a real and very visible tool—a pilot's sixth sense that elevates the professional flier above the unprofessional.

A certain pilot comes to mind when I talk about awareness and professionalism. He was a new customer to the fixed base operation and needed a check ride prior to renting the planes. We stepped out of the air-conditioned lobby and walked across the hot ramp to the Skyhawk's tie-down. As we neared the plane, he muttered to himself, "Compass must be leakin'." We reached the cockpit and looked in—he was right. He had connected a few simple facts. Before signing for the plane, he had reviewed the airframe and engine logs and noted that the plane had not been in the shop for several days. As the pilot approached the plane he caught a whiff of mineral spirits. He realized that two probable sources of the leak were a recently serviced engine or a worn diaphragm in the expansion chamber of the compass. A small thing indeed, but a visible mark of the professional nevertheless.

A student pilot provides another example of a flier building professionalism. The student had soloed two weeks earlier and

Preflight Planning and Professionalism

was preplanning a dual cross-country flight. She finished her inquiries and calculations and we headed for the airplane.

"Shouldn't be a bad crosswind when we land," she said.

"Okay . . . why do you say that?" I asked.

"Well," she said, "flight service gave the surface wind there as 12 at 30 degrees; the *Airport/Facility Directory* shows the runway 350 degrees. The handbook says the plane can take an 18-knot component and I've already landed with a 10-knot direct crosswind."

"Sounds great," I said. She had put together available information and evaluated it against her plane's ability and her own skill. She knew what she would likely face on landing even before we took off.

"Besides," she added, grinning, "if the wind does kick up, there's an airport eight miles beyond with three runways. OK, Teach?"

She had planned a possible "out." She was inexperienced in hours, maybe, but she was experienced in a good attitude toward her flying, most certainly.

The best part about total awareness and professionalism is that any pilot can develop it in his own flying. The first step toward understanding total awareness is to learn and practice total preflight planning. Many pilots do not plan their flight professionally simply because no one has shown them how. Once a pilot is shown all the factors that affect a flight and is told *how* and *why* he should evaluate those factors against his own skill and his plane's ability, I've found he is willing and able to adequately preplan his flights to virtually eliminate the prospect of pilot error.

The underlying goal of this book is to help the pilot develop a total and professional preflight plan that, at the same time, will help prevent pilot error. The book is structured around the three areas of preflight considerations: the pilot, the plane, and the environment. Each chapter presents an element of flight that calls for preflight preparation. Each chapter specifically identifies the factors within that element of preflight. The text explains why these factors are significant, how to research them, and finally how to evaluate them against pilot skill, aircraft capability, and the specific demands of the flight at hand. Each chapter is followed by a "review" section that is intended to be used as an on-the-spot guide to help plan actual flights while you further develop your preflight checks of yourself, your plane, and your sky.

RON FOWLER

Orlando, Florida

THE PILOT

Flying implies freedom to most people.
—Anne Morrow Lindbergh

CHAPTER 1: **Establishing Personal Weather Minimums**

RAGGED CLOUD BASES hung a leaden fifteen hundred feet above the rural northern Georgia airport. Rays of sunshine had tried several times to seep through occasional breaks in the overcast. But the next sweep of cold rain would again darken the glistening tarmac ramp area.

I had been sitting for two hours on one of the salvaged Volkswagen van bench seats that the fixed based operators or FBOs called "settees." A pilot and his family were also waiting in the airport lounge. They had made a precautionary landing just as the weather had closed in. Snatches of conversation had them homeward bound from vacation with home only another hour's flight southward. The two small sons had kept themselves entertained for a short while, but the fascinations of the red candy machine in the corner, the coverless magazines on the table, and the doodads in the showcase counter soon waned. The boys began asking about a time for going home: "Why can't Dad fly in the rain?" "Isn't it going to be dark soon?" As the weather-waiting lengthened, the mother looked more and more strained, and for the past hour the father/pilot had spent most of his time standing by the picture window, watching rain drip from the trailing edge of his Mooney's wings.

During a lull in the rattle of rain, a tenth stranger walked up to the pilot and asked the same question: "Still here?" I could almost hear that pilot's common sense snap. He bunched the muscles in his jaw, ran a hand over his face, let go an explosive sigh, and turned to collect his family. That's when I stood up to stick my oar in because I knew he had just come to a very, very bad decision.

One of the most important "firsts" a pilot must establish is his own weather minimums. He cannot let the opinions of others influence his decision of *go* or *no-go,* as did that pilot I described who was waiting out the weather in a northern Georgia airport.

Neither can a pilot rely on weather minimums of the Federal Aviation Regulations to provide guidelines of safety. One can witness this pilot error enacted at airports across the country on any foggy morning. There will be several flights flying under Visual Flight Rules (VFR) waiting for the fog to dissipate. Usually along about mid-morning the airport beacon shuts off, signifying that the field has just come up to the legal three miles. Immediately a few small planes will launch, and one has to wonder: Are those pilots capable of flying safely in three miles visibility? In most cases the answer is probably no.

I think that many pilots draw a false sense of security from the Federal Regulations. In many cases I believe their thinking goes something like this: "Surely the FAA wouldn't state the three miles visibility and a thousand-foot ceiling if they didn't think it was safe for me to use." This is erroneous thinking.

These VFR and IFR (Instrument Flight Rules) minimums prescribed by FARs rarely offer safe operating procedures for the average flight. It is not the intent of the FAA to do so. The regulations only prescribe absolute minimum weather conditions for legal operations. Each pilot must modify these legal minimums in light of his own experience, skill, and equipment if he is to fly safely. In most in-flight situations, for example, the average VFR pilot cannot operate safely when a scant but legal thousand-foot ceiling exists. Neither is the unpracticed IFR pilot safe in bringing his plane down to the published ILS (Instrument Landing System) minimum in most instrument conditions. Each pilot must establish his own personal weather minimums.

In order to establish your own cross-country minimums, you need to first identify the separate elements of aviation weather and consider how each affects the safety of your proposed flight. Basically, the VFR pilot must consider the minimums he is willing to tackle in terms of *reduced visibility, low clouds, extent of precipitation, intensity of turbulence,* and *strong crosswinds.* Let's examine these minimums in some detail.

Reduced Visibility

Four potentially deadly hazards can creep into the cockpit of a cross-country flight when reduced visibility keeps you from seeing far enough through the Plexiglas. First and foremost, reduced visibility hides any weather that lies ahead and around the airplane. More often than not, the pilot flies right into the weather before he sees it. Or if he does detect a darkening in time, he really doesn't know which way to turn, with reduced visibility on all sides.

We often hear the comment "The plane was caught by bad weather." I don't think any plane has ever been actually caught by weather—it doesn't reach out and make a snatch at a passing airplane. The pilot has to actually fly his plane into the weather. And this can easily happen if the pilot can't see where he's going.

The second potential hazard is getting lost over unfamiliar terrain, particularly when you cannot see landmarks six or seven miles ahead. And when you are lost your eyes tend to lock on

Reduced visibility hides any weather that lies ahead. The pilot can fly right into the IFR condition before he sees it.

Establishing Personal Weather Minimums 13

the chart in your hand, or the ground beneath your plane, so that you don't see the other planes. Getting lost also offers a good opportunity to run out of gas.

The third most apparent hazard of flying in restricted visibility is the increased chance for a midair collision as traffic and obstacles disappear into the murk. Two aircraft closing head-on at 160 knots, for example, cover five miles of flight visibility in less than fifty seconds. This may not give you enough time to spot the other plane coming out of the haze, become aware of the collision course, decide which way to duck, take action with the controls, and still have time for your plane to move safely out of the way.

Fourth and finally, reduced visibility often leads to disorientation. Visual references are often hard to discern. What we

---

HAZARDS ASSOCIATED WITH REDUCED VISABILITY

1. **WEATHER IS HIDDEN**

2. **CHANCES INCREASE FOR GETTING LOST**

3. **CHANCES INCREASE FOR MID-AIR COLLISION**

4. **MAY LEAD TO DISORIENTATION**

Knowing the hazards of poor visibility is the first step toward setting your own visibility minimums for cross-country flying.

think we see and what we really see in poor visibility are not always in agreement. Viewed through a veil of precip, for example, a sloping layer of cloud or smoke lying close to the hidden horizon line soon has you flying with a wing low. Or should you fly through five miles and haze and let a light shower hit your dirty windscreen you lose your visual references. You are suddenly engaged in on-the-job IFR training with only yourself aboard as instructor.

Knowing the hazards of poor visibility is just the first step in setting your own visibility minimums for cross-country flying. It is beneficial for a pilot to actually experience reduced visibility—not just imagine it from written words. The best way to achieve this is simply to take short flights around your familiar home area when the low visibility is known, with an instructor or another experienced pilot aboard. See for yourself what four, five, six, or seven miles visibility actually looks like. Gauge the degree of difficulty in comparing landmarks to your sectional and spotting traffic; imagine deteriorating weather a half mile beyond the farthest feature you can see on the ground; roll into a steep 360-degree turn to experience the tendency toward disorientation. Decide for yourself, from practical experience, just what minimum visibility you are willing to tackle.

FLYING IN MARGINAL VISIBILITY

Once you have decided upon the minimum visibility you are willing to accept, adopt a number of safeguards to monitor when flying in this marginally acceptable condition.

One, listen very carefully to the weather briefing, and if the low visibility is associated with low clouds or precip, know the cards are starting to stack up against you—IFR conditions can lie hidden just a few miles beyond your flight.

Two, even in dry skies keep a frequency tuned to Flight

Watch. This FAA facility (also called En Route Flight Advisory Service or EFAS) is manned by weather specialists and is capable of giving you up-to-the-minute weather that lies directly ahead. All you need do to receive the information is simply call on their common frequency of 122.0 and state your position. The proper station will respond and can usually contact any plane above three or four thousand feet.

Third, take extra preflight precautions against getting lost in poor visibility. The greatest hazard that stems from getting lost is the stress that it places on the pilot. And a stressed pilot very often acts in illogical haste, and more often than not makes the wrong decision. Whenever I think of the stress that a lost pilot feels, I think of a particular dual training cross-country that I flew with a pilot several years ago. He was not a beginner. He held his private certificate, had a hundred hours experience, and was working toward his commercial. Part of this training consisted of a three-hour dual flight across unfamiliar ground in far less than ideal VFR conditions. We took off under a 3,000-foot broken deck with six miles visibility and the knowledge that afternoon summer rains would meet us en route.

Two hours out, the ceiling turned overcast and dropped quickly to a lead-gray 2,000 feet. The few good landmarks that the unpopulated countryside offered began to disappear under the darkening woolly sky. The pilot then made a classic error— he deserted his straight heading and started to make shallow turns left and right toward the landmarks that did show up. I knew he would soon become lost. I kept my mouth shut but kept close tabs on our changing position and on the route to the nearby alternate airport that I knew we would soon need. A couple of signals confirmed the moment he realized he was lost. His feet began seesawing on the rudder pedals, and he started turning his sectional chart round and round, hoping that the answer might spring from it. It was then that we saw the curtains of rain out there at the limit of our visibility. The showers were staggered so that it looked like we were hemmed in on all sides. His anxiety grew and he tried to swallow it away. I still kept silent and let him get on with his job—and kept my eye on the airport that was now off to our right rear and just inside the veil of rain.

Then the overstressed pilot did a most unusual and illogical thing. Without really being aware of his action, I'm sure, he slowly brought the power to near-full throttle and started into a series of left 360 degrees—going nowhere but doing it fast.

Midway through the third circle I tapped him on the shoulder, pointed at his white-knuckled grip on the shoved-in throttle, made circles in the air with a forefinger, and turned an expression toward him that asked "Why?" Then I leaned over, told him that I knew exactly where we were, had an airport in sight, and would give him one minute on his own to put a plan together. With the stress diminished, he grinned and picked up his sectional chart. He quickly figured from his last known position and planted a thumb at our present position on the chart. Then he estimated a heading to the airdrome symbol, turned the plane, and saw the airport. Now he was a pilot who knew where he was and exactly where he was flying to. Every trace of nervousness disappeared and he made a smooth landing despite the gusty crosswind and an approaching heavy line of rain that already had the far end of the runway buried under a white froth as we touched down. He gave me a very graphic example of how simply getting lost can create stress that leads to such poor judgment.

A few preflight actions are helpful when tackling marginal visibility and substantially reduce the risk of getting lost. Preplan the simultaneous use of all three navigational methods: (1) pilotage (use of landmarks), (2) dead reckoning (use of known winds), and (3) radio (VOR and DME). In planning these take time to draw your course line on your sectional chart. Pencil the line heavily so that it won't disappear in a bouncing cockpit. Draw the line even if you are following an airway for those light blue airway lines all look alike when things get a bit hectic. Be sure to circle prominent landmarks every fifty to sixty miles, and note the estimated lapsed time from takeoff that you can expect to overfly them. When conditions are marginal, knowing that you have flown over a familiar landmark at the right time is a tremendous boost to a pilot's confidence.

Preplan a heading that compensates for the predicted wind. And plan to make use of any VOR that establishes a leg of your flight.

Once in the air and flying under marginal conditions, exercise all three methods of navigation simultaneously. Keep a minute-by-minute account of your flight along your penciled line with thumb and landmarks. Pay close attention to your preplanned heading. The most common error that gets pilots lost is the tendency to "wander" when the visibility diminishes. Strive also to keep the VOR needle in the "donut." Anytime one of the three navigational methods seems out of phase with the other two, find out why. If you have been holding a steady heading, for example, and the VOR and landmarks show you drifting off course, look into the reason. It could be that a nearby area of weather has shifted the wind or your Heading Indicator needs resetting. Running multiple navigational methods adds another measure of safety to your flight. Low visibility has the ability to take away part of your navigation on a moment's notice. Reduced visibility over sparsely landmarked terrain can easily cancel your sectional chart for several minutes at a time. This, however, is no great problem if you continue holding a good heading and VOR needle until the next landmark comes into view. At the same time, poor visibility often forces a pilot closer to the ground to maintain visual contact—possibly low enough to eliminate VOR navigation. This should be no great problem for the pilot who had been minding his pilotage and dead reckoning.

Keep in mind the cold fact of flying that pilots only see a third of their significant traffic, even in good visibility. Defensive flying is the answer to avoiding a midair collision in poor visibility. Be concerned with three basic defenses:

1. Increasing your efficiency in spotting traffic
2. Making yourself more visible to the other pilots
3. Avoiding unnecessary exposure to congested traffic

Take the first step toward better plane spotting while you preflight your airplane—clean the Plexiglas, inside and out. Remember, a full load of passengers mixed with the high humidity often associated with poor visibility added to an outside air temperature lower than the cockpit's dew point—and you get a quick smear of IFR. Some pilots take along a terry-cloth wipe; it's not a bad idea.

On those flights that tax your vision, put the passengers to work. Give them a preflight briefing and assign each a sector of sky to guard. Most passengers appreciate the responsibility, do a good job, and report traffic with the eagerness of a nervous B-17 tail gunner.

When visibility does drop, take advantage of the service Air Traffic Control (ATC) offers: VFR radar traffic advisories. This service is like having an extra set of eyes aboard, but you need to remember that ATC's prime responsibility is to IFR traffic. This means that they can aid VFR pilots on a "time permitting" basis only. If their IFR work load increases while they are giving VFR advisories, they may stop the service without advising you. Nevertheless, any VFR traffic advisories from the radar controller serves to increase your efficiency at spotting traffic.

At the same time, it is just as important to make yourself

Establishing Personal Weather Minimums

more visible to the other pilots during those flights into low visibility. Take the first steps during your walk-around preflight aircraft inspection. Make sure the landing light works and give the strobe a test run. Then, once aloft in the haze, fly with your landing light blazing away. Even if you don't see the head-on traffic, it is sure going to notice you! Strobes, of course, are very effective during hazy conditions, even in daylight. This cannot be said of the red rotating aircraft beacon. During a daylight flight the plane usually becomes visible before the beacon does. (If the haze is heavy enough to reflect the winking strobe, however, turn it off. Disorientation can result.)

Another valuable tool in flying marginal VFR is your aircraft transponder. Activate the VFR code on your transponder, even if you are not in touch with the controller. At least he will know of your presence and alert the traffic he is working.

It is so important to increase your own visibility anytime that you are approaching an uncontrolled airport. To remind ourselves of this, all we need do is simply remember that most midair collisions happen on the final approach to an airport without a control tower. To help the other pilots in the pattern spot you, give an accurate position report to unicom while you are still at least two minutes out. And plan to enter the downwind leg at a forty-five-degree angle. This is what the other pilots expect to see, and the forty-five-degree approach gives you a panoramic view of the pattern. If you see a nearby plane and even remotely suspect that the pilot does not see you, do two things immediately. One, announce your presence over the unicom frequency, and two, *roll your wings*. Not gentle little waggles, but crisp thirty-degree rolls that attract the attention of even the most unvigilant pilot. (Be sure to quickly advise your passengers what you are up to, or they will start taking a loud opinion poll!)

When flying in your minimum acceptable visibility, do everything in your power to avoid unnecessary exposure to heavy traffic. One way to do this is to choose your cruising altitude defensively. If you fly in excess of 3,000 AGL (above ground level), for example, you should adhere to the VFR cruising altitude rules of FAR 91.109. These rules have us flying the odd thousand-foot levels plus 500 feet (3,500, 5,500, 7,500) when on a magnetic course of 0 through 179 degrees. Even thousand-foot levels plus 500 feet (4,500, 6,500, 8,500) apply when flying a magnetic course of 180 through 359 degrees. This simple system provides a degree of aircraft separation. And enough pilots adhere to the rule to make it an effective deterrent to midair collision.

These rules, of course, do not apply to cruising altitudes you choose *below* 3,000 AGL. When flying at these lower levels, I use a defensive measure that you might find useful. Most pilots fly the *1,000-* or *500-*foot increments on their altimeters. When flying below 3,000 AGL, I hang my altimeter needle on the *250* or *750* mark; 2,250 or 2,750 instead of 2,500 or 3,000, for example. The few planes that have whizzed 250 feet above or below me have proven the practice worthwhile.

Another crowded traffic situation exists at VOR airway intersections. Many pilots navigate so that they avoid passing directly over these intersections during flights through restricted visibility. Also, it is often wise to avoid an ILS centerline below 3,000 feet within ten miles of the runway it serves because you can expect high speed IFR traffic turning and descending on final approach. Rightly or wrongly, the instrument pilot's attention is very likely to be glued to his panel, and he simply won't see you.

There may come a time in your flying when visibility drops to the point where you find yourself following a highway, river, power line, or railroad. I still like to remember that time as a fledgling commercial pilot, when we flew down to retrieve the

Let your altitude play a significant role in defensive flying. When cruising below 3,000 AGL, hold your altimeter on the *250* or *750* mark. Fly 1,250 instead of 1,000, for example. Most pilots fly the *1,000-* or *500-*foot increments on their altimeters.

hub and bent prop from a medium twin that had tried to land on a southern Florida duster's strip. We took off at mid-morning—Eddie, the veteran, leather-faced maintenance chief; his mechanic; the tool boxes; and I. We climbed into a hazy summer sky that was already gathering puffballs of cumulus. The plane was a big, strong cargo hauler that must have been designed by Fruehauf. We had pulled the passenger seats out of the long fuselage. Only the pilot's seat and the small aft bench seat for the mechanics remained, so I had the flight virtually to myself. Midway down the state those puffballs started to build rapidly, and the back of my mind told me that the return flight might have complications. But I was still at that point in flying where I thought that to show concern was to show a lack of enthusiasm. Besides, I had the controls of a super plane and was being *paid* to fly it.

We landed at the strip, took an hour in sweltering heat to remove the damaged prop, and hoisted it aboard as distant thunder was echoed from a blue-gray massing of cumulus to the north.

Shortly after takeoff I started detouring weather with wide turns and giving up altitude to stay out of the clouds. What began as scattered white rain squalls quickly gathered into a line of hard weather that kept forcing me westward. Yet common sense and a wind off the Gulf of Mexico told of even heavier weather waiting in that direction.

Respectable turbulence kept the stick and rudder bar busy. Constant turns and a sectional chart that kept bouncing off my lap soon had me wondering where we were. The bases of the clouds had already pushed us below VOR reception, when, over the darkening landmarkless terrain of the state's southern interior, visibility took a further plunge. Rain hit the windscreen like a fire hose and sent a bucket of precip roaring across the top of the cockpit. A slight turn pulled us out of the rain again,

Establishing Personal Weather Minimums

but I knew things were going to get worse. I had to find an airport fast—but which way? I was flying the bucking plane two-handed and deep into wondering what to do next when I felt rather than heard the clomp-clomp-clomp of Eddie coming up front for a look. He gripped the seat back, leaned past my right shoulder, and squinted into the grayish white air around us.

Eddie: "Lost, huh?"
Me: "Uh huh."
Eddie: "Not to worry—used to live down here . . . see that road?"
Me: "Uh huh."
Eddie: "Follow it to the Texaco station . . . then take a left along the power line to . . . careful! Don't move in too close to the road! . . . See! . . . See what I mean!"

A southbound Piper Cherokee flew over the road following our landmark at our altitude.

Eddie: "When you follow roads in this kind of stuff let the other guys fly the center line. Stay as far away as you can and still see. Now follow the power line and turn west on the highway . . . airport's only another ten minutes. I know the way . . . we'll make it."
Me: "Uh huh."

And so we picked our way through the ragged scud, Eddie leaning over my shoulder and pointing and guiding with a stubby finger, flickers of lightning reflecting off his age-lined face; me following and flying with uncertain "Uh huhs" all the way to touchdown.

When the visibility is low let the other guys fly along the road's centerline, sky-wise Eddie advised. In short, that flight was worth the lesson—not to mention other lessons that came to me from that flight. For instance, about listening to any doubts I had in the back of my mind. Or discovering that my voiced concern does not expel me from the cockpit. And I learned it is foolhardy to leave the airport behind with a suspicious sky ahead without a routine preflight call to Flight Service Station (FSS). (Even though the field did not have a phone, I'm sure the farmhouse down the road did. Or I could have circled the field to reception altitude and obtained my preflight briefing there.) And make a 180 back to clear air to sort things out, while there is still plenty of time to do so. Those are early lessons that I learned well. Oh, plane and sky still teach me new facts, but nowadays there are not as many on a single flight. I suppose I am finally getting it together.

Low Clouds

Whenever clouds force a cross-country flight to within 1,500 feet of the ground, the pilot, whether he knows it or not, is in trouble. That pilot has lost two reserves that mean so much to a flier's safety: abundant altitude and plenty of maneuvering room. Further, low clouds can easily push a plane below radio communication and navigation range, only thirty miles or so at 1,000 feet. At the same time, navigating by landmark becomes doubly difficult with the distorted perspective inherent in low altitude flying. As an additional hazard, a pilot must face wind changes in both direction and velocity as he descends from cruise altitude. (Surface wind usually shifts about forty-five degrees left of the wind experienced at normal light-plane cruising altitudes, and halves its velocity.)

Not only does low altitude flying complicate navigation, it also increases a pilot's chance of bumping into something hard. Low clouds, as I have pointed out, squeeze all VFR traffic down into a narrow band of airspace. Many of those pilots have their attention focused on charts and landmarks and coping with the difficulties of navigation, and not on seeing airplanes. Flying

Establishing Personal Weather Minimums 25

HAZARDS ASSOCIATED WITH LOW CLOUDS

1. NAVIGATION MAY BE DIFFICULT

2. AVOIDING TRAFFIC AND OBSTACLES MAY BE DIFFICULT

Before you establish your weather minimums, consider the two principle hazards of flying beneath low clouds.

1,500 feet or so above the ground is not so bad if the terrain is flat and free of tall obstructions. But on most lengthy cross-country flights you encounter both ridges and antennas. What may be 1,500 feet of clearance one minute can become impact the next.

Make personal experience a part of your preflight decision when you establish the minimum cloud deck you are willing to fly under on cross-country trips. With an instructor or another weather-experienced pilot aboard, set out on a short cross-country when the broken layer or overcast is of a known value and is low enough to plant concern in your mind. Plan this flight with caution—after all, you are setting out under marginal conditions. Then, carefully analyze the weather briefing to assure yourself that the clouds will probably drop no lower. To cover yourself on that "probably," select a training route sprinkled with alternate airports along the way. You will then discover firsthand the difficulties encountered in navigating close to the ground, such as spotting traffic and seeing obstacles.

As you set your minimum acceptable cloud height, consider

a value that lets you clear the cloud base by at least 1,000 feet as well as 1,500 feet above the terrain. I realize that the FARs require only 500 feet clearance beneath clouds when you fly below 10,000 MSL. But if you have ever had the silver-gray belly of a jumbo jet drop through the clouds to only 500 feet above you and a half mile ahead, you *know* that isn't enough clearance. (This is one of those experiences in life that provide a subject for conversation for weeks.) In addition, as you arrive at your personal weather minimum for low clouds, be sure to allow a margin for further cloud deterioration while en route cross-country.

When the clouds are down at my personal VFR preflight minimum, I take a few extra precautions before I get into the plane. My main concern is that the clouds may drop below my acceptable minimum once I am airborne. With this in mind, I route my flight to keep alternate airports within twenty minutes flying time of my flight path. In most areas of the country, airports are numerous enough to keep the detours acceptable. And if the extent of detouring is unacceptable, I simply delay the flight until conditions improve. But I can't justify taking off into marginal weather without giving myself a series of "outs."

Once my route beneath the clouds is decided, I plot the course on the chart with a heavy line. This anticipates a strong reliance on pilotage if the clouds should push my plane below VOR reception range. To further aid pilotage navigation, I circle prominent landmarks at twenty- to thirty-minute intervals along the route and jot down next to them the estimated lapsed time from departure. This lets me better monitor my progress and gives a number of intermediate goals as I fly through the marginal conditions. I choose landmarks that stand out even in the poor perspective of low-altitude flying. This means the landmarks must be large or tall—for example, towns, bodies of water, TV antennas, farms, or fire towers. Roads, small streams, and railways, I find, are hard to spot when the going gets low. As a hedge against any clouds lowering even further, I also circle any ridge or obstacle that rises within a thousand feet of my expected cruise altitude.

One modern-day hazard a pilot must face when traveling beneath low clouds is the military low-altitude training route. These military routes normally lead toward restricted areas, and the military jets fly these corridors in excess of three hundred knots within 1,500 feet of the ground. The big problem is the fact that you cannot see the jets soon enough to dodge them. If you anticipate having to cruise near 1,500 feet and your route lies within twenty miles of a restricted area, consult with FSS to see if there is an active low-altitude training route associated with it. If there is, you would be wise to detour your low flying elsewhere. A few years ago, I conducted wildlife survey flights around the perimeter of a restricted area that served a training route. After a few close whooshes, I traveled to the military base to talk to a flying officer who was involved in the flights. I told him of my inability to see his planes in time and waited for a suggestion. He told me that we were even—the fighter pilots can't see the small planes either!

When the clouds are low and you receive your weather briefing from FSS, pay extra attention to three factors:

1. Are the clouds expected to get lower? If so stick to your acceptable preflight minimums.

2. Is there a second marginal condition waiting for you—reduced visibility, rain, turbulence? If so, think twice about going.

3. Is your arrival time near sunset? If so, be prepared for an early darkness under the low clouds.

Once aloft, you must recognize the fact that low clouds place you and your plane at a disadvantage. This calls for a one hundred percent effort on your part. Minimize the distractions of conversation and sight-seeing. You have a full-time job at hand—delivering plane and passengers safely to the destination airport.

It takes concentration and hard work to keep a minute-by-minute track of your current position. It means keeping a thumb along your course line as landmarks drift by while you are equally busy flying the plane accurately, managing communications, evaluating the surrounding weather, constantly updating your alternate courses of action, and looking out for traffic in the limited airspace beneath the clouds. But you *must* keep track of your exact position. The moment you become uncertain of where you are, things tend to go wrong; those clouds start to come down. And racing clouds to the ground is no fun if you don't know which way it is to the finish line.

As you fly cross-country under low clouds, maintain a weather briefing with FSS or Flight Watch. It is certainly reasonable to call at twenty-minute intervals to update any weather changes at your destination airport and en route to alternate airports.

Alternate airports are your best insurance when you fly in marginal conditions. In addition to having an alternate in mind, be sure to mentally update the estimated heading and time to it every few minutes. Whenever possible, use *paved* airports as your alternates; sod fields are difficult to find when you are flying low. If the clouds force you to within 1,500 feet of the ground, turn toward your nearest alternate at that time, and tell FSS of your plan; any further cloud lowering is going to put you in a serious position.

By the time a VFR pilot finds himself flying within a thousand feet to stay out of clouds, he should be at his alternate airport and in a position to land. But it may sometimes happen that a pilot finds himself below a thousand feet with no runway in sight. Poor judgment, bad information, or the capriciousness of weather itself can throw plane and pilot into that predicament. For whatever reason, the pilot who finds himself skimming the terrain under dark and threatening clouds knows without being told that he is in trouble. And he feels the ring of alarm that tells him he must now deliver some of the best airmanship he has ever flown.

A number of rules of thumb may help the pilot as he presses closer to the ground. He first must settle down to the job at hand and not let the alarm of the emergency grow into the illogical panic of desperation. He can find some reassurance in knowing that clouds rarely come right down to the ground, so long as he maneuvers to stay clear of any rain.

Ridges and obstacles, however, can easily rise to close the gap between cloud and collision. The pilot's first defense against such a collision is simply to slow down. A speed equal to 1.5 times the flaps-up stall speed (lower green-arc speed) is appropriate in most light singles. Things happen quickly at low altitude. The slower speed often lets a moment's inattention pass without harm; it gives the pilot extra reaction time as well as better maneuverability by virtue of a shorter turning radius.

The pilot flying within a thousand feet of the ground should turn his auxiliary fuel pump on, switch to the fullest tank, and lower his flaps to the first notch. A malfunction in the engine-driven fuel pump can cause a momentary loss of power that a pilot just cannot afford at minimum altitude. Switching to the fat tank at the outset of the low flight can prevent a busy pilot's forgetfulness from turning into an accident. One notch of flaps increases the margin between flying speed and stalling speed during any quick turn away from an obstacle or ridge, yet doesn't greatly affect the plane's performance.

A low-flying pilot should keep his hand on the throttle and add a touch of power anytime he turns away from a hazard. This reduces the plane's stall speed during the brief period of a higher load factor. He should run with his carburetor heat full hot. The high humidity of low clouds and a slow-turning engine are very apt to produce carburetor ice. Frequent power changes keep him from detecting it early, and his first warning might be a sputtering engine.

A pilot forced to fly low should stay in constant communication. The problem here is selecting a station in range (about twenty-five miles at five hundred feet with the average light plane radio). His best bet is to use the emergency frequency, 121.5. While his sectional chart may not show a primary facility within hearing distance, chances are pretty good that an ATC remote facility is nearby, and that the emergency frequency will reach it. The people on the ground cannot fly the plane or make decisions. But they can provide information to a pilot who is quickly running out of options.

When preplanning a VFR cross-country flight, the question often arises, "Should I plan to fly beneath the cloud layer or go VFR over the top?" While there can be no hard-and-fast answer to this, it is normally safer to fly beneath the cover. This assumes, of course, that the bases meet your personal weather minimum for low clouds. If your minimum is not met, you are better advised to delay or postpone rather than go VFR on top.

If a VFR pilot decides for some reason that he must plan the flight above the clouds, he—along with the FSS weather briefer—must carefully weigh the risks. First, are the reported clouds no greater than "scattered," with no second layer above them? Will the bases permit safe VFR beneath? Then, does the forecast indicate either an improvement or no change, and are the destination airport and an alternate expected to be cloud-free an hour before to an hour after the expected time of arrival? Finally, the pilot should get a *current* pilot report. Chances are good that FSS is in communication with a plane on the proposed route. And if requested, the FSS briefer will contact that plane for a report.

Once in the air, the pilot must reevaluate the decision to climb above the clouds. While still beneath, he must determine the extent of coverage and look for a second, higher layer. Unless those clouds are widely scattered (less than a quarter of the sky covered) with no additional clouds above, he should stay beneath. It is often difficult to estimate cloud coverage from the air. A pilot beneath the clouds cannot see the size of the open areas. But a look downward toward the cloud shadows on the ground tells the story. Before climbing up through the clouds, he should take a final look ahead for any darkening that foretells of significant clouds a little farther along the route.

From the moment the pilot levels off above the clouds, he should maintain radio contact with FSS and keep in tune with any changes in the weather, behind as well as ahead, to cover a possible retreat. It is difficult to visually determine any change in the cloud cover that lies ahead. The perspective at only a thousand feet or so above the clouds is the problem. Nevertheless, the pilot must look for buildups that spell increased coverage ahead and keep close tabs on the clouds immediately beneath the plane.

Any thickening of clouds is reason for an immediate descent. No pilot should continue above closing clouds hoping for an improvement "just ahead." Any hesitation could find him trapped. I have seen widely scattered clouds become an overcast in less than a minute. It can happen more quickly than the time it takes to get the plane down through the gathering cover.

If the pilot should find himself caught above the clouds, he should take three immediate steps to prepare an escape. First, he should mark his position and time on his chart as the last landmark disappears. This lets him keep a running estimate of position as he later ticks off attained positions along the course line at ten-minute intervals. (A common mistake is a failure to maintain the course. It is tempting to turn the plane as you search for a way down. It is far better, however, to hold the plane steady and let your head do the swiveling.)

Second, the pilot should throttle back to conserve fuel. A ceiling can cover an extensive area and remain there for a considerable time.

And third, he should quickly alert FSS of his predicament. Together with the FSS personnel, a decision can be made as to the best direction to fly. Then the pilot would be wise to contact ATC for vectors to the VFR letdown point.

The pilot caught above the clouds faces three common sources of unavoidable instrument conditions. Picture yourself as a passenger in the plane. FSS advises that the nearest reported VFR lies two hundred miles from your position. They also advise of a strong head wind from that direction—and the fuel gauges are

---

HAZARDS ASSOCIATED WITH VFR ON TOP

1. CLOUDS MAY EXTEND BEYOND THE PLANE'S RANGE

2. PLANE MAY BE CAUGHT BETWEEN CONVERGING LAYERS

3. BUILDING CLOUDS MAY OUTCLIMB AIRPLANE

The pilot caught above the clouds faces three common sources of unavoidable instrument conditions.

solidly below half. Or picture the plane between layers. With ground and horizon lost and the jaws of those layers converging, vertigo is likely—with loss of control an almost certainty. Or finally, picture the clouds building upward. The pilot climbs to stay out of them, but it is easy to see that they are rapidly outclimbing the plane—with the plane already nearing its service ceiling.

Whatever the source, the IFR specter usually gives the pilot several minutes to comprehend the inevitable loss of VFR references. And it is at this point that we leave the pilot, gathering himself for the terrible decisions and actions to come. Because the uncertain and unproven procedures for extracting a VFR pilot from IFR conditions are beyond the scope of this book; they are a subject in themselves.

Extent of Precipitation

The average pilot has little trouble picking his way around rain or thundershowers, as long as they cover less than half the area he is flying through. If you add a little more precipitation to bring that coverage up to fifty-fifty, his wings won't stay dry. Once the ground area surrounding the flight is half covered with precip, it becomes very difficult to remain VFR or to avoid turbulence. The problem is not solely the percentage of land covered; a pilot certainly doesn't take up half the sky with his plane. The problem with navigating around precip is that it rarely stands still for us. Areas of rain easily move thirty miles an hour. Oftentimes two adjacent showers move at different speeds, and their exact direction of travel is hard to detect from the cockpit. So—to use an analogy—given sufficient coverage and movement, the pilot rides a pinball, trying to stay in the clear, waiting for weather to push him where he doesn't want to fly.

Navigating through an area of rain activity is a subject generally glossed over in our student training days. That is why I feel that a pilot must go out and experience the problem, with a weather-wise pilot aboard and over familiar ground. Consult with FSS just before your training flight so you know the areal coverage and can compare the weather briefer's terms (widely scattered, scattered, broken, solid) with what you see through the windscreen. Then decide after the flight the extent of coverage that you would feel comfortable with over unfamiliar territory, and make that value your personal weather minimum for precip.

If you plan to fly cross-country on a rainy day, review the latest radar summary chart with the weather briefer. This chart shows the areas of precip, the intensity of the rain, whether it is decreasing or increasing, and the percentage of the area actually under rain. These charts are updated frequently when there is activity, but the character of precip can change quickly. For that reason see if you can coax a current pilot report from the mists of the rainy area you want to fly through. As in dealing with any factor of marginal VFR, if the weather briefing indicates a second factor, think twice before you put the plane in the air. For example, scattered rain and either reduced visibility or low clouds constitute a good excuse to delay the flight until the sky improves. Poor visibility lets you fly right into the rain, and a low deck blinds you to buildups several miles ahead.

Before you put the plane into the air, draw your true course on the chart and make certain there are alternate airports within the area of rain. Even though you can expect the rain to detour you away from your course line, you want to know your position relative to that line and to a nearby airport as you pick your way through the weather.

Once aloft and approaching the area of precip, you must reevaluate the weather picture with a windscreen inspection. Does the rain still hide less than your allowable percentage of terrain? Are the visibility and cloud cover (outside the rain itself) still

Establishing Personal Weather Minimums

within your acceptable standards? And is turbulence—before you reach the rain—still nominal?

If the final decision is to penetrate the area, the pilot should take three defensive steps prior to making any detour. First, make sure ample fuel remains. Running the tanks down to the quarter mark with heavy rain on all sides has cost many a pilot the friendship of his passenger. Second, check the outside air temperature. A pilot can fly through light rain if visibility is good enough for him to see the horizon through it, and the temperature is well above freezing. And third, establish and maintain communication with FSS. Report the weather conditions to FSS, and keep updated on any changes that lie ahead and at your alternate airports.

Let five rules of thumb guide your actions as you begin to detour around rain or thunderstorms:

Rule 1. *Establish your exact position before each detour.* If you start detouring from an uncertain position, you are lost before you even begin. Further, the rain that hides landmarks will not let a pilot put his act together easily while he detours.

Rule 2. *Plan your detour route before you turn away from your present heading.* Don't arbitrarily turn left or right and then decide where you want to go; this practically guarantees a lost pilot. While holding your heading, plan your detour as a straight leg that clears the rain to a landmark on the chart. Before turning, pencil the detour on the chart, and estimate a heading and time en route to the landmark. Now you know where you're flying to and how you're going to get there.

Rule 3. *Keep FSS posted on your detour plans.* Let them know your detour route and when you expect to turn back to your planned flight route, and give FSS a revised ETA to destination.

Rule 4. *During the detour, maintain an awareness of your*

position relative to your primary course and relative to a nearby alternate airport. You will soon want to return to your planned route, and you may need that alternate.

Rule 5. *At the end of the detour, return to your original course.* Plan a straight-leg return to your planned flight course. Keep in mind that the inability to return to course is often a pilot's early warning that the weather is controlling his moves.

When working your way between rains, stay alert for any deterioration in conditions: dropping visibility, developing low clouds, or spreading rain. When the air is saturated IFR comes quickly. The moment that the rain coverage exceeds your acceptable percentage, head for an airport. And if you feel any uncertainties about navigating, get in touch with ATC for radar assistance.

A pilot should be at his alternate airport by the time the ground is half covered with precip. If he waits until that time to head for his alternate, he cannot navigate in straight legs. His flight becomes a series of sweeping turns just to stay VFR. At this stage visibility between the rain deteriorates and low clouds develop. The pilot *needs* radar assistance which may, or may not, keep him clear of instrument conditions.

Intensity of Turbulence

Pilots think of turbulence in terms of passenger discomfort, aircraft control, and (gulp) structural damage. It doesn't take much bouncing to discomfort most passengers. The intensity the weather briefer reports as light or moderate is usually enough to do the job. If the passengers should become, ah . . . err . . . umm . . . discomforted, then be forewarned that they hold a terrible sword of retribution. Let's, for the moment, look at the practical subject of barf bags. Do not depend too heavily on those little plastic bags you buy at the airport counter. By the time the passenger's fumbling fingers fish the flimsy thing from its envelope, he probably has already, ah . . . eased his discomfort. It's a good idea on bumpy days to drop by the grocery store on the way to the airport. Get a bunch of bread-loaf size brown paper bags. Write BARF on each and scatter them around the cockpit. Now that's a barf bag! You may have read some theories about things to say, or things to do, to keep an anxious passenger from, ah . . . discomforting up. Forget what you've read—none of it works. A passenger considers being air-sick his personal decision. If he decides to, he will, so don't try to interfere.

If the flight looks choppy, insist that the passenger take some motion-sickness tablets before he climbs aboard. Some passengers refuse as they feel it an unheroic act. Don't hesitate in these stubborn cases to push the passenger to the ground, plant a knee on his chest, hold his nose till he opens wide, and poke those pills down his gullet. This is for his own good; it preserves the tranquility of the cockpit and is within your scope of authority as pilot in command.

When turbulence is reported as severe, you can expect to have trouble keeping the plane under positive control. It usually causes large changes in airspeed, heading, and altitude. While this in itself is not dangerous, a pilot's real concern is knowing that severe turbulence is normally punctuated with blasts of extreme turbulence.

Extreme turbulence is another matter. The criteria for this intensity make it likely there will be structural damage, and an airplane that is violently tossed about is practically impossible to control. The unfortunate pilot who flies into extreme turbulence will hear the price his plane is paying in pops, groans, and snaps. He will have decided, even before he lands, that it is now time to trade it in on a newer one.

Most pilots who have fought severe or extreme turbulence want no part of another round. The best way to avoid these degrees of turbulence is with a good weather briefing and knowledge of when you can expect to encounter the conditions en route. Thunderstorms, mountain ridges, and strong winds are three common sources facing light planes.

Any thunderstorm that is producing continuous lightning, hail (recognized from the cockpit as a greenish cast in the clouds), or "Mama" bases, is also producing severe or extreme turbulence. This turbulence may easily extend ten miles from the cloud at altitudes of 5,000 feet or more. At the lower altitudes the plow wind of the storm collides with the surrounding air to produce the hazardous turbulence for five to ten miles. Inside or beneath the storm, turbulence is disastrous. A VFR pilot can easily avoid thunderstorms if he has reasonable visibility and the surrounding rain leaves him plenty of maneuvering room.

If winds at the mountain ridge level are in excess of fifty knots, severe or extreme turbulence over that ridge is guaranteed. The best bet is to detour around the mountain or delay your flight until the winds subside. Do not attempt to go through the passes or valleys if the winds over the top are strong. The mountain tops and ridges can funnel the wind so that velocities are *increased* in the lower valleys. However, shift the wind direction and those same tops and ridges can protect a valley or pass from high winds. The wind at a valley airport, for example, could be light, yet at the tops and ridges, turbulent. Be forewarned as you leave the valley airport to climb well above the ridge level while you stay over the valley and well away from those mountainsides.

If your flight path approaches a mountain ridge from downwind, your plane can succumb to the ocean of air tumbling over and down that ridge. To prevent this, plan to clear the ridge by 3,000 feet. Begin your climb fifty miles from the ridge to avoid the downward tumble of onrushing air. Then make a final evaluation of the turbulence as you approach the ridge. If you are having no trouble controlling your airplane, you can probably cross the ridge. But if the turbulence is giving you trouble even before you reach the ridge, turn around.

Surface winds in excess of forty knots usually spell severe or extreme turbulence aloft. You are practically certain to encounter significant turbulence at the lower altitudes if this wind flow is associated with a front, or if there is a temperature inversion. These strong surface winds normally preclude a takeoff anyway, but the problem arises when the wind picks up after lift-off.

Anytime a pilot encounters severe turbulence en route, he should take four immediate defensive steps.

1. *Slow down.* You slow a car down to hit a bump. You do the same thing with a plane for the same reason—to keep from shaking it to pieces. In a car the slower you approach the bump, the better. This is not so with a plane, however. The vertically moving turbulent air can momentarily increase the wing's angle of attack as it blows upward across the leading edge. If that plane is moving too slowly it may stall. The correct method is to pick the fastest speed that the plane can fly without suffering damage by an abrupt move. This speed is called *maneuvering speed* and is found in the aircraft flight manual. You may have time to look it up, and you may not. If not, just look at the airspeed indicator and double the speed printed at the low end of the green arc; it is close enough to do the job. Note the power setting it takes to establish your maneuvering speed. If the turbulence worsens, the airspeed indicator may lose its reliability in the eddying wind. If the instrument does become erratic, fly the constant power setting.

2. *Cinch your seat belt and shoulder harness down tight.* Severe turbulence can throw your head into the cockpit's metalwork hard enough to knock you cold. And if you are in my age

bracket (between forty and reward), your neck will snap with a pop that gives a week's worth of headaches. Advise your passengers to tighten their belts and secure any loose objects within the cockpit. A whizzing camera, for example, makes a fine cannonball when the going gets extreme.

3. *Maintain attitude rather than altitude.* Your altitude fluctuates widely in severe turbulence. Let it, as you probably have the whole sky to yourself anyway. Control pressures that maintain an altitude also place extra stress on the plane. Just try to hold the plane as level as possible and let it ride the waves. If you have an autopilot, be sure to disengage the altitude hold mode.

4. *Contact FSS.* Tell FSS what's going on and see if they can find you a safer altitude or a turbulence-free area. If a change in course is mandated, turn slowly to minimize the aerodynamic load on an already stressed airplane.

Strong Crosswinds

The botched crosswind landing is the single leading cause of aircraft accidents. Not many people get hurt, but planes do, and a tattered wing tip is expensive to replace. If the crosswind touchdown is rough enough to collapse a gear and put the prop into the pavement, the resulting pilot error will cost the price of a new Ford.

A flier's ability to safely handle a particular crosswind component depends primarily on two factors: pilot skill and aircraft capability. The ability to land safely in a crosswind is a pilot skill that quickly erodes. There are two reasons for this. The crosswind landing is a complex maneuver. Few maneuvers match it in terms of timing and coordination. Maintaining this skill, means constantly using it, yet the pilot does not often put his skill to use; he wisely keeps it on the ground when the winds are high. As a result, most pilots have a problem with handling crosswinds. One rule says use it or lose it. Another rule says don't use it when it will do the most good, that is, when it tests your skill.

The most practical solution to this dilemma lies in dual instruction. When you look out the window and the wind is blowing harder than you would want to land in, call your favorite instructor for an on-the-spot schedule. Then with the instructor aboard, resharpen your skill and determine your personal crosswind limitation. The prudent pilot conducts this recurrent training flight every few months. It is the best insurance against a crosswind accident that he can buy.

The airplane's crosswind capability is usually stated in the aircraft manual. If not, you should estimate it as one-third of the plane's stall speed. Then be sure to select the lower of your skill or the plane's capability as the maximum crosswind component you are willing to tackle.

To include this crosswind component in your preflight planning, just compare the destination's forecasted surface wind with the runway headings at that airport. The FSS weather briefer will give you the wind, and you can find the runway heading in the Instrument Approach Procedure Charts that are on display at FSS.

Once you know the wind and the runway heading, apply some yardsticks to determine the crosswind component:

1. If the wind lies within thirty degrees of runway alignment, estimate the crosswind component at one-third the wind's velocity.

2. If the wind lies from thirty to sixty degrees across the runway, estimate the crosswind component at two-thirds the wind's velocity.

3. If the wind lies from sixty to ninety degrees across

the runway, estimate the crosswind component to equal the velocity.

As you evaluate the crosswind during your preflight planning, keep two things in mind. First, the wind could change by the time you arrive. If the wind is unusually high and the airport destination doesn't have a choice of runways, make sure a nearby alternate does. Second, remember that wind directions (except for those given by the tower) are stated in *true* and runway headings are stated in *magnetic*. If your destination lies in an area of significant variation, be sure to convert direction so that you are comparing apples to apples.

Night Flying Preflight Weather Minimums

Cross-country night flight is a personal pilot decision. There is no doubt that night flight presents a greater hazard than daytime flying. The reason is simple—we can't see as well at night. The same holds true for automobile driving, doesn't it? Yet many people drive at night. Night flying comes down to a question of acceptable risk; and it is a pilot's personal decision based on his own skill, knowledge, and judgment.

A pilot setting up his night flying preflight weather minimums would do well to review the factors that seem to repeat themselves in night accident after night accident. Heavy weather is not normally in the nighttime VFR accident report. Most pilots know better than to set out with steady rain, low clouds, or poor visibility. Typically, the pilot of the ill-fated night flight is non–instrument rated. In examining reports, one finds that the flight was conducted in reasonable night visibility but there were widely scattered showers in the vicinity. One usually finds, also, that the flight went down in a brief shower over an unpopulated area. In sum, the pilot had good visibility, only scattered show-

---

COMMON FACTORS IN NIGHT-FLYING ACCIDENTS

1. PILOT NOT IFR RATED

2. SCATTERED SHOWERS IN VICINITY

3. DARK AND UNPOPULATED AREA

A pilot setting up his preflight night-flying weather minimums would do well to review the factors that are common to most night accidents.

ers, and an unpopulated terrain. This is great weather for daytime flying, but these factors are a different matter at night. The problem is that the pilot cannot see the rain showers or clouds in time to avoid them, particularly over unlit countryside. The light rain that he can easily see through in daylight reduces his nighttime forward visibility with instant IFR. With no lights immediately beneath the pilot for even the meagerest sort of reference plus the light turbulence of the rain, vertigo and loss of aircraft control are probable.

The night-flying VFR pilot, then, should modify his daylight preflight weather minimums for night flights. He may very well want to double his standards for visibility and cloud height. He might take the precaution of routing the flight to avoid uninhabited, dark countryside. It is not unreasonable for a pilot to decide against night flying when even scattered precip or light turbulence is forecast. In addition, a pilot's crosswind landing capability diminishes somewhat at night. A pilot does, after all, depend on quick visual references during the round out to detect drift. In daylight these visual clues are all around him. At night visual references are confined to that egg-shaped patch of a plane's landing light.

A pilot's decision to fly at night and his nighttime weather criteria are individual matters. All depend on the degree of experience he brings to the cockpit.

To those pilots inexperienced in night flying but contemplating nighttime cross-country trips, my advice is simple. Get several hours of dual cross-country training over both familiar and unfamiliar territory at night. Be familiar with the problems. Then evaluate your experience and decide on your own night flying preflight weather minimums.

IFR Weather Minimums

In addition to the VFR weather factors, the IFR pilot must also consider the elements of *takeoff minimums, landing minimums,* and *icing conditions.*

TAKEOFF MINIMUMS

The FARs do not mandate takeoff minimums for private flying. Yet most experienced IFR pilots want enough ceiling and visibility at takeoff to allow an instrument approach back to the airport. A malfunctioning engine, in-flight fire, or sudden encounter with unforecast conditions may require a quick return to the ground.

LANDING MINIMUMS

The minimum descent altitudes or decision heights and the forward visibility prescribed by a published approach procedure are not adequate for many IFR pilots. The published minimums do not guarantee safety. They only state the absolute minimum conditions that a pilot may legally fly in, consistent with his skill, equipment, and flight situation.

ICING CONDITIONS

Icing intensities are reported in terms of trace, light, moderate, and severe. Each intensity describes a rate of accumulation and the specific hazard to flight.

Trace—ice becomes perceptible. Deicing or anti-icing equipment may not be required for exposure under one hour.

Light—rate of accumulation may create a problem on exposure exceeding one hour, unless deicing or anti-icing equipment is employed.

Moderate—the rate of accumulation creates a hazard on even short exposure. Deicing/anti-icing equipment or diversion is necessary.

Severe—neither deicing nor anti-icing equipment can control the rate of accumulation. Immediate diversion is necessary.

Conclusion

Once you have considered the hazards within the various elements of aviation weather, establish your own personal preflight weather minimums. Be sure that your personal minimums allow for some furthur deterioration in the weather to occur after you have left the vicinity of your departure airport. If, for example, you decide you don't want to navigate beneath clouds of less than 2,000 AGL, you might decide that you won't take off unless you have cloud bases of at least 2,500 AGL. Then if en route clouds start to lower to 2,000, you have time to head for your alternate en route airport.

Similarly, if you decide that five miles visibility is your minimum for safe flight, then you may wish to peg your departure minimum at seven miles. If you feel that moderate turbulence is the maximum bouncing you want, then depart only when the forecasts call for no more than light turbulence. IFR pilots might want to add a certain percentage to the published minimums for their personal approach minimums.

Treat each element of aviation weather with the full realization that in-flight conditions can worsen, and allow yourself a margin of safety.

Once you have established your personal weather minimums, reduce them to writing. The time to establish your personal minimums is *now*. Do not put it off until you are faced with an impending flight.

Your own guidelines, in writing, will help reinforce preflight decisions concerning marginal weather conditions. Without a doubt, reference to your written guidelines greatly offsets the "let's get going" pressures that often push a pilot into marginal conditions that logic tells him are beyond the capability of his equipment or his skill.

In Review

- Each pilot must firmly establish his own weather minimums; he cannot let the opinion of others influence his go/no-go decision.
- The weather minimums of the FARs do not normally provide guidelines of safety—only legal minimums.
- Basically, the VFR pilot must consider the minimums he is willing to tackle in terms of reduced visibility, low clouds, extent of precip, intensity of turbulence, and strong crosswinds.
- Visibility below five or six miles hides any weather that lies ahead and around the airplane.
- It is easy to get lost over unfamiliar terrain when you cannot see landmarks six or seven miles ahead—which in turn offers a good opportunity to run out of gas.
- The most apparent hazard of flying in restricted visibility is the increased chance for a midair collision. Two aircraft closing head-on at 160 knots cover five miles of visibility in less than fifty seconds.
- Reduced visibility often creates visual illusions and leads to disorientation.
- It is beneficial for a pilot to experience reduced visibility, not just imagine it from written words.
- On flights into reduced visibility, put the passengers to work; assign each a sector of sky to guard.
- Take full advantage of VFR traffic advisories when visibility drops.
- Fly with your landing light on in poor visibility. That way, even if you do not see the head-on traffic, it is certainly going to see you.
- Fly the VFR cruising altitudes when above 3,000 AGL. Under 3,000 AGL consider flying the 250-foot increments (2,250 or 2,750).

- When you follow highways and roads in poor visibility, stay as high as you can and still see the road.
- Anytime clouds force a cross-country pilot to within 1,500 feet of the ground, he is in trouble. The pilot has lost two reserves that are vital to safety: abundant altitude and ample maneuvering room.
- Low clouds complicate navigation. A low-flying pilot has a distorted perspective of his landmarks, and he may be beneath VOR reception range.
- Low clouds push all the VFR traffic down into a narrow band of airspace creating a potential for midair collisions.
- Make personal experience a part of your decision when you establish your personal minimums for low clouds.
- Consider a personal weather minimum that lets you clear the terrain by 1,500 feet or more, and the cloud bases by 1,000 feet.
- When planning a flight beneath low clouds, choose a route with alternate airports within twenty minutes reach.
- Navigating beneath low clouds demands that you know your exact position all along the route. Furthur deterioration may require a fast alternate arrival.
- By the time a pilot finds himself within 1,000 feet of the ground, to stay out of clouds he should be at his alternate airport in a position to land.
- A pilot planning VFR over a cloud layer greater than "widely scattered" runs the risk of getting caught on top.
- The average pilot has little trouble picking his way around rain as long as it covers less than half the terrain. Precip reaching fifty-fifty usually prevents flight through the area.
- In dealing with any factor of marginal weather, beware of the existence of a second weather factor. For example, scattered rain *plus* reduced visibility or low clouds calls for delaying the flight until the sky improves.

- Let five rules guide you as you detour around rain or thunderstorms:

    1. Determine your exact position before beginning each detour.
    2. Plan your detour route before you leave your present heading.
    3. Keep FSS posted on your detour.
    4. During the detour, remain aware of your position relative to your original course and the closest alternate airport.
    5. At the end of each detour leg, return to your original course.

- A pilot should have his plane on the ground by the time the terrain is half covered by precip.
- A pilot should think of turbulence in terms of passenger discomfort, aircraft control, and structural damage.
- Turbulence defined as light means occupants may feel a slight strain against seat belts.
- Turbulence defined as moderate means unsecured objects move about and occupants feel a definite strain against seat belts.
- Turbulence defined as severe means the aircraft may be momentarily out of control and occupants are thrown violently against seat belts.
- Turbulence defined as extreme means that the plane is impossible to control and structural damage may occur.
- Thunderstorms, mountain ridges, and strong winds are three common sources of significant turbulence.
- Moderate to extreme turbulence may extend ten miles from a thunderstorm.
- If winds at the mountain ridge level exceed fifty knots, severe or extreme turbulence is almost a certainty.

- If you have no trouble controlling your plane as you approach a ridge, you can probably cross it. If you are having control problems before you reach a ridge, turn back.
- Surface winds in excess of forty knots usually spell significant turbulence aloft.
- Whenever a pilot encounters severe turbulence en route, he should take four immediate defensive steps:

    1. Slow to maneuvering speed.
    2. Cinch seat belts and harnesses down tight.
    3. Maintain attitude rather than altitude.
    4. Contact FSS to locate smoother air.

- The inablity to handle the crosswind landing is the single leading cause of aircraft accidents.
- The ability to land safely in a crosswind is a pilot skill that quickly erodes. Recurrent training is the answer.
- If the airplane's crosswind capability is not stated in the aircraft manual, estimate it as one-third of the plane's stall speed.
- Rules-of-thumb to estimate the crosswind component:

    Wind within thirty degrees of runway alignment—estimate crosswind component at one-third of wind velocity.

    Wind thirty to sixty degrees across runway—estimate the crosswind component at two-thirds the wind's velocity.

    Wind sixty to ninety degrees across the runway—estimate the crosswind component to equal the wind's velocity.

- Three factors are common to night VFR accidents:

    1. The pilot is not IFR rated.
    2. There are scattered light showers in area.
    3. The area is an unpopulated terrain.

- Before deciding to fly solo at night, get several hours of night cross-country training.
- Once you have considered the hazards within the various elements of aviation weather, establish your own personal weather minimums and write them in your logbook.

Preflight Aids

A. PERSONAL WEATHER MINIMUMS FOR VFR FLIGHT

Reported Visibility _____

Cloud Bases _____

Max. Wind Velocity _____

Max. Extent of Precip. _____

Degree of Turbulence _____

Thunderstorm Activity _____

Night Flight Notes _____

_____

B. PERSONAL WEATHER MINIMUMS FOR IFR FLIGHT

Take off Min. : Vis. ____ Ceiling _____

Extent of Icing _____

Degree of Turbulence _____

Thunderstorm Activity _____

Personal Minimum Altitude
for Approaches _____

Personal Minimum Visibility
for Approaches _____

C. OTHER PERSONAL WEATHER CONSIDERATIONS _____

_____

The time to establish your preflight weather minimums is *now*, not later when the flight is at hand. Once you have considered the hazards within each element of aviation weather, decide on personal weather minimums that are consistent with your skill and experience and put them in writing.

CHAPTER 2: **Gauging Your Accident Potential**

PILOT ERROR doesn't just happen; it is a pilot's deliberate creation. In one instance, a pilot arriving on downwind radioed the tower he had just run out of gas. He safely made his emergency landing, and our tug went out to tow him in. I recognized the shaken pilot as he climbed from the cockpit ahead of his wife and small son. I knew he had flown in from Atlanta and had overflown an airport thirty miles northwest. The fuel gauges had to have been bouncing on empty even then. When he had a few moments, I walked over to him and asked, "Why?" He answered quite honestly: "Because my wife's folks were waiting for us, and I didn't want to run late and worry them."

Every year good logical-thinking pilots come to grief because they suddenly act in an illogical, inexplicable manner. It's easy, as we hear the account or read the report, to think that we would never do such a thing. But in truth we all hold within us a real potential for tragedy—a potential time bomb.

Put simply, it is a matter of misplaced priority. It is so absolutely easy for us to momentarily forget piloting's number one goal of flight—the safety of the passengers. In almost every instance of pilot error that I've seen, the pilot has temporarily displaced this top priority with an earth-bound consideration of lesser value. The moment this happens, that pilot is reaching for an accident; the time bomb starts ticking.

Pilots' illogical actions seem to stem from six basic human frailties. Let's look at some examples and watch the priorities bounce.

Vanity / Pride / Illogical Considerations

The two pilots were ferrying their single-engine planes back from Texas in close formation. Each pilot was well experienced and thoroughly familiar with the high performance machine he flew.

Mid-afternoon found them climbing eastward on Vector 20 to stay above the cumulus gathering about them; their tanks were big and they could file IFR if need be. The Texas border, passing down through Sabin Lake, drifted behind them unseen beneath the broken cloud deck. There was good VFR above the clouds although maybe a little choppy. But that small amount of turbulence only let them better feel the two hundred knots they were clicking off. The clouds pushed higher and they talked it over on 122.9—whether to stick together VFR or file and go their separate IFR ways. The decided VFR.

Towering cumulus punched up through the overcast, bringing rain to the hamlets hidden far below. The thunderheads were far apart with clear, cool air between them. Shallow turns detoured them neatly along the airway. The pilots thought it a good flight.

Two miles above the Louisiana lowlands, something happened that is difficult to explain. While looking at one another through their Plexiglas canopies (I'm sure) and with mike buttons unpressed, the two experienced pilots deliberately flew, wing tip to wing tip, into a mature thunderstorm. Once inside the dark nine-mile-high storm, there was little they could do to hold back the inevitable. Wind shear and hail pulled the bucking planes apart, even as the pilots asked themselves "Why did we do it?"

SIX SOURCES OF PILOT ERROR

1. **MISPLACED VANITY**

2. **PHYSICAL CONDITION OR MENTAL OUTLOOK BELOW PAR**

3. **A PILOT IN A HURRY**

4. **DEFIANCE OF THE SYSTEM**

5. **A PILOT'S BELIEF THAT IT CAN'T HAPPEN TO HIM**

6. **GET-HOME-ITIS**

Pilot error doesn't just happen, but is usually a pilot's deliberate creation. It is very easy for a pilot to momentarily forget his number one priority—the safety of his passengers.

Many months later, an airline captain and his copilot taxied their jet onto the runway, surrounded by swirling snow and bitter cold. As they straddled the centerline, they calmly discussed their iced-up wings and then launched into the wet, sticky, frozen sky they knew would not admit them. And in those sinking seconds before tearing impact even as they quickly acted to avert what they knew was about to happen, I'm certain each asked himself "Why did we?"

The only answer I can come up with as to why these tragic

pilot errors occurred is *vanity/pride/illogical considerations*. I have to put all three in the same category because they seem to interrelate; I cannot tell where one mental attitude ends and the next begins. I have seen the clues, though. This phenomenon most commonly happens when two pilots are at the controls in the same plane when the opportunity for a really bad pilot error comes before them. Then each refuses to verbally question what he *supposes* the other has decided, or perhaps hesitates to be the first to back away from that supposed decision. But even one pilot at the controls can do the job quite nicely. The pilot I described earlier who decided to ask the impossible of his plane rather than worry his in-laws is a good example.

We are all susceptible to vanity and pride for these are human traits. Used properly, they may add to our enjoyment of life, but used improperly in the cockpit they become an in-flight hazard. The best alternative I can suggest is to ask ourselves a question as we preplan the flight: "Will this action I'm proposing add to the safety of the flight, or am I acting simply out of regard for someone's personal feelings?"

Physical Condition and Mental Outlook

All of us have those days when we just can't get things right. If you have trouble just parking your car in the FBOs lot, for goodness' sake don't go flying that day. Similarly, don't climb into the plane on that day you "feel like you're coming down with something." Many pilots do, with the idea that their health will improve with altitude. But it won't; in fact it will probably get worse. Imagine coming down with the initial distress of flu and the destination still twenty minutes ahead!

We live in a pressure-cooker world. For most, a certain degree of stress is a normal part of daily life that does not deter us from our daily rounds. But stress can come in hard-to-swallow doses. And each of us has a capacity for stress.

Take a preflight peek into your stress bucket to see if it is nearing the brim. Look for any physical signs that tell you your bucket is full: stomachache, tensed muscles, shaking hands, or a headache—only *you* know your own symptoms. If you are nearing your capacity, don't go flying. Chances are an in-flight stressful situation could run your bucket over, with chances then good for a colossal pilot error. (A few tensed-up pilots even try to *use* a plane as a tranquilizer, which is pretty dangerous medicine.)

In addition, look toward your mental state that day. A flaring temper can easily push a pilot beyond safety, just to "show 'em who's in charge" anyway. Further, temporary depression or anxiety is another especially grim ghost to turn loose in the cockpit. Either can create a "What's the use?" outlook. As a result, if a critical situation calls for instantaneous action, an anxious or depressed state of mind could delay the pilot's response to the point of low-grade suicide.

Fatigue and hunger are often underrated causes of pilot error. We do not make sound business or personal decisions when we are overtired, and we shouldn't expect our in-flight decisions to be any different. Our preflight planning must accommodate our own personal endurance with rest stops.

Hunger affects different people in different degrees. Some persons can go two days on a celery stalk while others can't function properly without a steak-and-eggs breakfast behind them (I personally start kicking at small dogs and children if I've missed lunch). Plan your flight to include chow stops or a brown bag to satisfy your own needs.

In a Hurry

Boy, don't we make mistakes when we rush ourselves? And those mistakes are not just confined to rushing madly through a job. A lot of rush activity can set up mistakes that keep on occurring well after things have seemingly settled down.

Take my very first charter flight, for instance. I was at home mowing grass when the call came from the FBO I was flying for. The boss wanted me to fly him and his family to a resort two hundred miles up the coast. They were waiting at the plane . . . could I get right down? I grabbed a quick shower and made a fast call to FSS. The weather was closing in; you will have to "hump it," they said. I jumped in the car, hit the starter, and discovered I had a dead battery. I ran to a neighbor and asked for a push. All the time I had one eye on the clouding sky. I got the car running and swung onto the boulevard. I then started fighting six miles of bumpers to the airport. I had to keep the motor revved so it wouldn't die at a light. The clouds were gaining on me. I even ran the caution light.

As I swung into the airport I saw the boss loading the last suitcase and Mrs. Boss sitting in the back seat . . . hurry! I tried to make the preflight look thorough; I started the engine and made a fast taxi to the runway. Wouldn't you know, I was number six for takeoff? I noted that the low clouds were moving up from the southwest. We were flying north so I knew I had to move it! How close can I keep the prop to that guy's rudder? Scud started rolling over the airport; drops came on the windscreen. I knew I would be IFR in seconds. Finally I scooted down the centerline and was off and climbing northeast toward sunshine. I thought, "Now I can relax and start being a full-time pilot."

Twenty minutes later I was into the flight watching the beach drift by, and all of a sudden saw that ground *was* only drifting

Gauging Your Accident Potential 59

by! I looked at the airspeed. It showed twenty knots low and . . . oh good grief! The gear was *still down and locked!* I had to keep a poker face and decide what to do. It was not simple sneaking that gear up. It lifted with a great long, shiny chrome bar. To work it you had to throw that bar through a 180-degree arc. To make sure it held, you needed to move the chrome bar with a flourish that would make a drum major blush—all accompanied by a clatter and clank that sounded like a tramp steamer hoisting anchor. There was simply no way to sneak all that. I had the option to pull up the gear, arrive on time, and admit all, or I could squint steely-eyed ahead and complain loudly of an unforecast head wind.

Rushed preflight planning often sets up pilot errors that don't manifest themselves, just as in my horrendous experience, until later in the flight. My point here, of course, is to allow ample time for planning. If a flight's immediacy will not allow time to safely plan, exert the willpower to postpone the trip—it is just another case of priority.

It is not unreasonable for a trip of a few hundred miles to require an hour's planning and preflight activities: checking weather, gauging aircraft performance, calculating weight and balance, laying out the route, researching the destination airport, selecting alternates, looking up frequencies—it all takes time.

If the nature of your flying seems to crowd adequate preflight planning, put a number of time-savers into play. A kit made up of your preflight materials and kept at ready not only saves time but adds orderliness to your planning. Such a kit might include:

Sectional charts for your area of operations
Low-altitude IFR en route charts/instrument approach charts
Airport/Facility Directory for your section of the country
Airman's Information Manual

FARs, Parts 61 and 91
Plotter and computer
Pilot logbook
Aircraft operating handbook with performance tables
Photocopies of your aircraft's weight and balance data and loading tables
Aircraft checklists

Many pilots keep their preflight planning materials in their airplane. But if time saved is worth money to you, consider keeping a second preflight planning kit in your home or office. You often have the opportunity to start your planning well in advance of the flight.

If you are the sole operator of the plane you fly, consider conducting the aircraft inspection after you land, or on a day between flights. Then only a brief surface inspection is needed before the flight to detect any exterior damage that could come to a parked plane.

If you fly repeated routes, retain a written record of true courses, distances, airports, and radio frequencies. These items are not subject to frequent change, and a written record already has much of your planning accomplished. (More on this in Chapters 8, 9, and 10.)

Defiance of Government Regulations

Let's face it. Many persons today harbor an antagonism toward governmental agencies and bureaus. In recent years we have felt the weight of heartless, greedy, and illogical bureaucratic action that seems to stem our efforts to simply enjoy a live-and-let-live existence. This is bound to cause a feeling of resentment toward anything governmental or bureaucratic. But in the interest of air safety and the continued efficiency of aviation, we cannot look with contempt on the procedures and regulations governing our lives aloft. And let me say that it is also unjustified to do so. Of all the government agencies, the FAA is certainly among the least bureaucratic. This is simply because most of the agency's individual personnel really care about aviation. Most are pilots and enjoy the sky quite apart from their tasks. Insofar as procedures and rules of flight are concerned, I cannot think of one that doesn't make sense in the cockpit. But then they should make sense; they are the result of joint effort between the administrative personnel (who are pilots) and the users (also pilots).

Antagonism against government regulations in the air is often a prelude to senseless disaster. It is a mood that lets a pilot view a weather briefing with contempt, or causes a pilot to delay an ATC demand for immediate action. If a pilot suspects that he is carrying a grudge against the FAA into the cockpit, he should remind himself of three facts:

1. Antagonism leads to senseless accidents.
2. The guys he thinks he's getting even with are in a steel and concrete building on the ground.
3. He is the one hanging in the sky in a flimsy airplane.

Defiance of the system can come from another quarter: the old pro who feels that his skill and experience elevates him above the rules. Unfortunately, that feeling doesn't fit him in too well with all the other pilots, and we must classify him as an in-flight hazard.

"It Won't Happen to Me" Syndrome

An airplane is designed to operate within a very narrow performance range. Weight, stress, aerodynamic ability, and aircraft performance are all unyielding values that make no allowances for the wishes of people. There is no such thing as a "forgiving" airplane. Load any plane a fraction out of balance, slightly exceed the design load factors, fly the wing a half degree past its critical angle, or ask it to stay airborne one minute beyond its endurance, and the flight will likely fail.

Similarly, a pilot is often reaching toward pilot error when he chooses to take on a critical flight situation that demands skill beyond his demonstrated ability. By this I do not mean that a pilot shouldn't expand his skills. He should—but in a controlled situation. A pilot with mediocre short field ability, for instance, can spend many hours sharpening his landing skill on the middle 2,000 feet of a long runway. But that same pilot would be ill-advised to take his mediocre skill into a *real* 2,000-foot strip. There is a popular feeling among pilots that faced with a truly critical situation we will react beyond our normal ability. But this is rarely true. It has been my observation that a pilot really behind the eight ball often reacts with far less than his normal skill and judgment.

Get-Home-Itis

No doubt for many pilots, the homeward flight is often a "go at any cost" affair. I have noticed a common thread running through these instances of poor judgment. Often the pilot feels pressed to go because someone has arranged to meet him at the airport. For pilots who find themselves susceptible to get-home-itis, I suggest this: Don't arrange to have someone waiting for your arrival; call them after you get on the ground.

Human frailties, as I have outlined, often lead good pilots into illogical actions. What flier needs bad weather or a malfunctioning airplane for enemies when he has these human problems within? While the frailties may seem a bit nebulous, the results are often only too real. And every one of us is susceptible. As a final preflight defense against them, let me make a final suggestion for you to consider: Anytime you feel outside pressures pushing you into an action that defies your logic, leave the airport; you will be surprised how quickly the pressure dissolves with the waiting plane no longer in sight.

In Review

- Pilot error doesn't just happen; it is a pilot's deliberate action.
- Pilot error is usually a matter of misplaced priority.
- The pilot's number one priority in flight is the safety of the passengers.
- A pilot's illogical actions seem to stem from six human frailties:

    Vanity/pride/illogical considerations. This phenomenon most commonly causes error when there are two pilots at the controls.

    Physical condition and mental outlook. A temporary illness will probably worsen once a pilot is airborne. And if you are nearing your capacity for stress, don't go flying.

    In a hurry. A really good rush can set up errors that keep occurring well after things have seemingly settled down. If a flight's immediacy won't allow time to safely plan, exert the willpower to postpone the trip—it is just another case of priority.

    Defiance of government regulations. In the interest of safety and the continued efficiency of aviation, we cannot look with contempt on the procedures and regulations governing our lives aloft.

    "It won't happen to me" syndrome. There is no such thing as a forgiving airplane. Operate any plane a fraction out of its very narrow performance range, and the flight will likely fail. And a pilot facing a critical situation cannot be counted upon to react beyond his normal ability.

    Get-home-itis. Often a pilot feels pressed to go because someone has arranged to meet him at the airport.

- Anytime you feel outside pressures pushing you into an illogical action, leave the airport; the pressure dissolves with the waiting plane no longer in sight.

Preflight Aids

Personal Preflight Questionnaire

- Is every preflight decision based on fact and free of vanity?
- Is your physical condition at its best and free of any symptoms that signal possible impairment at altitude?
- Has an adequate time elapsed since the consumption of any alcohol, in order to prevent a reoccurrence of symptoms at altitude?
- Are you beginning the flight free of fatigue?
- Are you beginning the flight with your personal stress-load well below its capacity?
- Have you allowed adequate time for an unhurried preflight preparation?
- Is every preflight decision in accordance with established procedure and regulation?
- Are the anticipated demands of the flight within your demonstrated capability?
- Is the flight totally free of the "I'm going, regardless" syndome?
- Unless each question can be answered with an honest and unqualified "yes," the flight has a high accident potential.

A pilot must preflight himself as well as his aircraft and environment.

CHAPTER 3: Complying with FARs

WHILE COMPLIANCE with the Federal Aviation Regulations (FARs) does not guarantee safety, noncompliance is usually hazardous. The pilot who has satisfied FAR 61.57 with three landings over the past three months, for example, does not assure safety to his passengers. Noncompliance with the minimum recent flight experience requirements, however, certainly invites disaster to climb into the cockpit.

Liability and civil action are further reasons that should convince pilots that it is important to comply with the FARs. Most liability and hull insurance contracts are nullified by violation of specified FARs. Further, any pilot facing civil action arising out of an aircraft incident feels more at ease in court if he has not burdened his attorney with a fractured regulation. These facts are foremost in the mind of a taxiing pilot who has, for instance, just let a moment's inattention damage the wing tip of a chocked corporate Lear jet.

Nine FARs that pertain to the pilot are listed below. An initial reading of the regulations in their entirety is in order. (Every pilot should have current editions of Parts 61 and 91 of the FARs in his library of flying materials. These are ordinarily obtainable at fixed base operators.) I highly recommend using this abbreviated list of regulation titles and comments as a preflight reminder.

FAR 61.3 REQUIREMENTS FOR CERTIFICATES

A pilot must have his pilot and medical certificates with him when he flies; he must be prepared to show them to an FAA or NTSB inspector if asked.

Photocopies are not valid, and FAR 61.29 explains how to get a replacement copy if yours is lost or destroyed.

FAR 61.23 DURATION OF MEDICAL CERTIFICATES

A third-class medical certificate is valid until the last day of the twenty-fourth month following your medical exam. A second-class certificate reverts to a third-class certificate the last day of the twelfth month following its issuance.

FAR 61.31 GENERAL LIMITATIONS

Paragraph (e) of this regulation requires a logbook endorsement of competency from a flight instructor (subject to an exception) before the pilot is eligible to fly a high performance airplane.

A high performance airplane is one that has either more than two hundred horsepower; or retractable gear, flaps, and a controllable propeller.

FAR 61.51 PILOT LOGBOOKS

A pilot must log the hours spent in pursuit of a certificate or rating, or he must log the hours needed to comply with the recent flight experience required by FAR 61.31 and 61.57. No other flight time need be logged.

A student pilot must be the sole occupant of the plane to record "solo" time. "Dual" hours credited toward a pilot's certificate must be from a certified flight instructor. This does not mean that a student cannot fly under the guidance of a rated pilot who is not an instructor. Indeed, hours flown with a friend can be the most meaningful to a student's training. It is just that these hours do not count toward the certificate. (A word of advice here: Let the rated pilot do the takeoffs and landings.)

A pilot may log as instrument flight only that time that he flies by reference to instruments, in either actual or simulated IFR. If simulated, the log entry must show the name of the safety pilot.

FAR 61.53 OPERATIONS DURING MEDICAL DEFICIENCY

This is a regulation that pilots should be careful to observe. It tells us we must ground ourselves if we develop a condition that would keep us from passing a medical. The problem is that we are not medical examiners. If you suffer an injury or illness that even you suspect might qualify as a "medical deficiency," the safe thing to do is to see your medical examiner immediately.

FAR 61.57 RECENT FLIGHT EXPERIENCE

This regulation spells out four items of flight experience that must appear in the flier's logbook if he is to act as pilot in command of an aircraft:

Flight review
General experience
Night experience
Instrument experience

Let's look at each of the elements that are considered necessary to qualify as "recent flight experience."

Flight review. In order to fly as a pilot in command, a pilot must have successfully completed a biennial flight review within the preceding twenty-four months. A satisfactory flight test for a certificate or rating, however, serves as a flight review.

The review itself is conducted by a flight instructor of the pilot's choosing, and in an aircraft in which he is rated. The review is conducted in two parts: (1) a ground session and (2) an in-flight evaluation. It is the instructor's responsibility by regulation to review the current operating rules of FAR Part 91. But the pilot should make the review much more than that. The day preceding the review, the pilot should make a list of aeronautical knowledge items that he would like to review. The proper use of recently designated airspace, for example, or how to safely off-load fuel. I shall not forget the pilot who asked to review the loading procedure for his plane. He had purchased the plane several months before and just couldn't figure out the format of the weight and balance tables. Up to this point he had only flown the plane locally with a friend aboard. He and his wife had a vacation flight planned with another couple later in the week of his review. I asked him to write down the fuel he planned to carry, the estimated passenger weight, and as best he could guess, the luggage weight. He handed over his sheet showing full tanks, four people weighing 635 pounds, and (he guessed) 80 pounds of luggage. It was a combination that he had carried in his former plane many times but this plane was bigger. Together, we worked over the figures, through his plane's weight and balance tables, and found the plane was well over gross and loaded quite tail-heavy. He was surprised. We both walked away from that flight review feeling that FAR 61.57 had earned its keep.

It works best if you make the in-flight portion of the review a fifty-fifty partnership venture between you and the instructor. If you do, you will find that the flight becomes one of instruction as well as one of evaluation. Discuss the content of the flight before you take off, and design it for the type of flying that you expect to experience in the coming months. If you anticipate instrument operations, for example, ask the instructor to include an approach or holding pattern. If a flight to the mountains is in the offing, include a short-field landing with a crosswind. Make the review reflect your needs; the instructor can still easily evaluate your basic flying technique.

When the instructor records the satisfactory performance in your logbook, you're set for another twenty-four months. But if the instructor was *not* satisfied with your basic flying, he makes no entry in your log. Then you have some decisions to make. If you think the instructor made a bad call, you can see another instructor for another review, and there is no ''unsatisfactory'' endorsement in your log to influence his judgment. But if you agree with the instructor that your flying was below flight test standards, ask him for specific recommendations for improvement. If it is just a matter of rusty skills, go out and sharpen them by yourself to prepare for a rerun on the review. (If your prior review has expired, you need a rated pilot on board to serve as pilot in command.) An honest self-appraisal may show you that you initially lack the fundamental flying skills. If so, start your recurrent training and know that the flight review has done its job.

General experience. A pilot planning to carry passengers must have made at least three takeoffs and landings within the preceding ninety days, in the same *class* airplane that he intends to use. (Examples of class are single-engine land and multi-engine land.) If the plane is a tail wheeler, qualified landings are those that come to a full stop. You may, of course, comply through solo flight.

Night experience. The same rule applies. Three night takeoffs and landings within ninety days are mandatory if passengers are going to be aboard at night. Night is defined as one hour after

Complying with FARs

sunset to one hour before sunrise. Full stop landings are required, regardless of where that third wheel is hooked on.

Instrument experience. An instrument pilot cannot file an IFR flight plan unless he has logged six hours of flying on the gauges in the past six months. This flight time can be actual or simulated IFR and must include six instrument approaches. A flight simulator can serve for three of the six hours.

A noncurrent pilot cannot conduct IFR operations until he passes an instrument competency check, normally given by an instrument flight instructor.

FAR 61.118 PRIVATE PILOT LIMITATIONS

I think almost every private pilot knows that he cannot receive compensation for flying a plane unless it is in connection with his regular employment and is just incidental to the business at hand. What many private pilots may not know, however, is that regulation prevents the passengers from covering a pilot's expenses for the flight. The regulation *does* allow the pilot to share operating expenses.

FAR 91.3 RESPONSIBILITY AND AUTHORITY OF THE PILOT IN COMMAND

This regulation spells out role and obligations of "the pilot in command"; it places the full responsibility and the final authority squarely on the pilot during any flight.

What does "full responsibility" mean? In essence, this phrase means that the pilot is liable for anything that goes wrong during the flight. It is the flier's role to gauge the flying task, weigh the risks, and make the final decisions. The pilot is fully responsible, for example, for the mechanic who handed him a defective plane or the weather briefer who gave him the wrong information.

What is implied by "final authority," the other obligation covered by this regulation? This is the more difficult part of the rule to enact. It gives the pilot that "final authority" to do anything he has to do to react to a situation that may arise during flight that requires immediate action such as turning left rather than right to avoid head-on traffic. As every experienced pilot knows, the obligation of bearing "final authority" is one that is always foremost in the veteran pilot's mind.

FAR 91.11 LIQUOR AND DRUGS

Eight hours is required between the bottle and the throttle. But is this enough time? I think it is probably not, in most cases. Each can of beer, every four ounces of wine, or each ounce of liquor remains in the blood for three hours at sea level. At 5,000 feet, however, the time needed for sobering up is doubled. I am sure this is why a consistently high alcohol-related fatal aircraft accident rate continues. According to the latest figures, nearly one-third of all fatal accidents are alcohol related. I am further sure that most of those pilots did not weave their way into the cockpit. Most likely they had met the required regulation; however, when they climbed to high altitude they set their body clocks back some hours. As a result, the effects of the alcohol were still in their systems and they were not really "sober." Black coffee or cold showers do not help the situation. These remedies only produce wide-awake drunks. Time is the only cure—it should be twelve to twenty-four hours depending on the quantity of alcohol imbibed.

This regulation further states that a pilot cannot even carry a passenger who is obviously under the influence of alcohol. Why, they're trying to take *all* the fun out of flying, aren't they? Nothing in aviation affords the entertainment value of a chuckling drunken passenger who decides to help with the trim wheel as you flare out or starts fiddling with the tank selector handle just after takeoff.

FAR 91.11 prohibits a pilot from flying while using *any* drug that adversely affects his flying skill or judgment. It is advisable, if you are taking a prescription medicine, to call your medical examiner. He can check to see if the use of your prescribed medicine actually prohibits flight. He can also advise you what side effects may be encountered at different altitudes. If you have been taking a patent medicine and have had no bad effects with it in the air, go ahead—take a swig and fly. But don't change brands or try something new just before you take to the air.

CHAPTER 4: **Keeping the Pilot Flight-Ready**

THERE IS A SEGMENT of preflight preparation that every professionally oriented pilot must conduct on a continuous, progressive basis—keeping himself flight-ready in terms of basic pilot skills and aeronautical knowledge. This ongoing preflight program must perform two functions. First, the program must provide a schedule of "updater" training and study to prevent a pilot's attained skill and knowledge from deteriorating. Second, it must provide a means whereby the pilot can constantly monitor and evaluate the level of his skill and knowledge.

Obviously this preflight preparation cannot be accomplished in the minutes or even hours before the planned flight begins. A pilot, for example, cannot simply "shelve" his attained skill or knowledge and expect either one to be flight-ready at some future date. Both skill and knowledge have limited shelf life. They will remain intact only with regular practice and study.

This segment of preflight preparation is often overlooked by pilots. Yet it is just as vital to the safety of the flight as is any other part of a pilot's total preflight preparation. A careful pilot, for example, would not ignore a maintenance schedule on his airplane simply because the machine was once in satisfactory condition. Nor would he eliminate his preflight aircraft inspection just because his plane is well maintained. Well, the professional pilot feels the same way about his skill and knowledge.

Just because his skill and knowledge were once in satisfactory condition, he does not ignore the need for a continuing updater schedule to maintain his level of proficiency. He knows that he must also "inspect" this proficiency on a continuous basis to insure that his basic pilot skills and aeronautical knowledge remain in a constant state of preflight readiness. Here are some thoughts to help put together such a program.

Keeping Basic Pilot Skills Preflight-Ready

Basic pilot skills will remain preflight-ready only through a program of regular practice. This regularly conducted practice is easy to come by if the pilot flies on a daily basis. The facts of general aviation, however, are that most pilots do *not* fly on a regular basis. Most pilots consider themselves lucky if they get in three or four dozen flights in a year. And it is this majority of pilots who need a definite continuous program that will maintain their skill to a constant preflight readiness level. It takes a conscious effort toward such a program to obtain the maximum benefit from the irregular flights that they do make.

I think it is important for a pilot to realize that the frequency of flying is so much more critical to retention of skill than the hours of flying. The pilot, for instance, who makes a thirty-minute flight each week can expect to retain more of his basic skills than can a pilot making only a single four- or five-hour flight per month.

My own flying is an example of this "frequency of flight" versus "hours of flight." As an instructor I fly about a hundred hours a month. Yet when I return to the airport after a two-week vacation, I can easily see that my skills have eroded. Instead of being two jumps ahead of any student errors, I am a half beat behind. It then usually takes me a couple of flights before I am back up to par. A program to maintain a pilot's skill level must be designed to compensate for the average pilot's infrequency of flying.

Not only must this program be designed to maintain a pilot's level of skill, it must also provide a means to evaluate those skills on a continuous basis. Again, my own flying provides an example of a need for this means of evaluation. As I pointed out it is easy for me to see the deterioration that a two-week layoff causes in my flying. The deterioration is easy for me to see simply because I have a most visible test to gauge my performance by—my reaction to my students' errors.

The average pilot does not have such a highly visible gauge to evaluate his own flying against unless he deliberately creates one. The program we will discuss in this chapter does just that; it provides a visible means whereby the pilot may constantly evaluate his own level of basic pilot skills.

This continuous program of preflight readiness is designed to perform three primary functions:

1. It makes the best possible use of an infrequent schedule of flying.
2. It enables a pilot to maintain his flying skill even with infrequent flying.
3. It provides a method of continuous evaluation to insure a constant state of preflight preparedness.

The program consists of two separate phases. First, it is designed to make every flight a practice flight. Second, it includes a monthly "recurrent training" flight that tests the pilot's skill against the demands of two specific precision maneuvers. Here is how each phase works.

Make Every Flight a Practice Flight

Nearly every flight offers the pilot an opportunity to practice his basic pilot skills. This opportunity, however, does not come to the pilot uninvited. He must deliberately plan his flight as a series of practice drills. Each drill can be based on the skills needed to perform the different segments of a normal flight, from the takeoff to the landing. To turn a segment of the flight into a practice drill, the pilot needs to do three things:

1. He needs to decide the tolerances he is willing to accept as a measure of satisfactory performance for each segment of the flight. These tolerances are usually stated in terms of altitude control, heading control, and airspeed control.

2. The pilot must then reduce his tolerances to writing. The tolerances for each segment of the flight, stated on a three-by-five-inch card and stuck to the instrument panel, becomes a means of evaluation that cannot be easily ignored.

3. After the pilot has decided on his criteria for satisfactory performance and reduced this criteria to a visible test of his skill, he must endeavor to make his plane perform within the stated tolerances. In this way, the pilot turns each segment of his flight into a practice drill.

Here are some thoughts on several segments of a normal flight, and how a pilot may turn each into a drill that insures his flight readiness when it is time to fly. The tolerances stated for altitudes, headings, and airspeeds are only my suggestions. Each pilot must determine for himself the accuracy he wishes to maintain.

TAKEOFF

Begin the takeoff run from exactly astride the runway centerline to start the flight as perfectly as possible.

Keep the nose wheel on the centerline throughout the takeoff run. This requires a touch of right rudder with the initial application of power to counteract the "slipstream effect," and an additional application of right rudder to oppose "P-factor" (the propeller's left-turning tendency) at lift-off. Use the ailerons

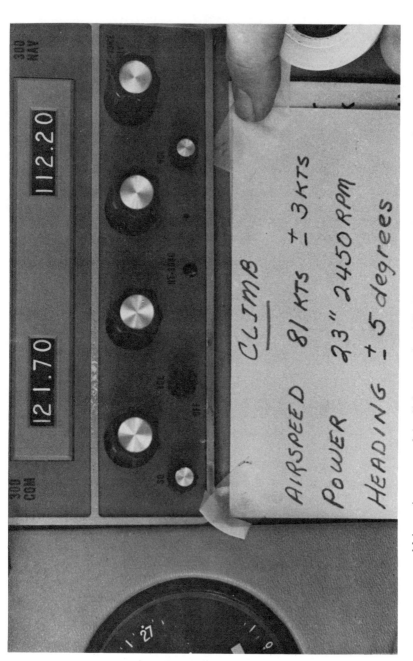

Make each segment of the flight a practice drill. Your tolerance for acceptable performance, written on a three-by-five card and stuck to the instrument panel, becomes a means of evaluation that is not easily ignored.

Keeping the Pilot Flight-Ready

to correct for any crosswind from the moment the plane starts rolling.

Rotate within three knots of the plane's recommended lift-off speed. This prevents either an excessively fast ground run that causes undue wear and tear to the plane, or an early lift-off that invites a stall.

Keep the initial climb-out directly over the centerline. This requires an immediate wind correction angle at lift-off and a continued right rudder pressure against the continuing effects of P-factor.

CLIMB TO ALTITUDE

Climb to altitude within three knots of manufacturer's recommended best-rate-of-climb airspeed. This degree of accuracy means that the pilot knows exactly how high to hold the nose relative to the horizon. Important, also, is proper trimming.

Maintain a climb heading within five degrees of the climb heading desired. Either adjust rudder trim or hold right rudder against the effects of "torque."

Maintain the exact climb power setting recommended by the manufacturer. It will be necessary to advance the throttle at each 500-foot interval in order to maintain a constant manifold pressure.

LEVEL-OFF FROM CLIMB

Lead the level-off from climb by 20 feet for every 100 FPM on the vertical speed indicator.

Maintain climb power throughout the level-off. Reduce the throttle to cruise power just as the airspeed indicator reaches cruise speed. This assures the best possible acceleration to cruise speed while preventing a loss of altitude during the final moments of level-off.

Maintain the climb heading throughout the level-off. This requires the pilot to slowly reduce right rudder pressure as the plane accelerates from climb speed to cruise speed.

CRUISE

For cruise maintain the exact power setting recommended by the manufacturer for the percent of power desired.

Maintain the precise fuel flow required by the cruise power setting. Remember that fuel flow must be readjusted with any change in altitude, power, or carburetor heat.

Maintain the heading within three degrees of that desired. If a slight change in heading is needed to recenter the VOR needle, do not arbitrarily turn left or right toward the needle. Instead, turn to a specific compass point.

Maintain the cruising altitude within 50 feet of that desired. This degree of accuracy depends upon a properly trimmed airplane, a constant awareness of pitch attitude, and a constant power setting. An error in one or more of these three factors will cause a large variance in altitude.

DESCENT FROM CRUISING ALTITUDE

In descent from cruising altitude maintain the manufacturer's recommended descent power setting to prevent overrevving or rapid cooling. Adjust the fuel flow every 1,000 feet as required.

Maintain the descent heading within five degrees. Use rudder trim or rudder pressure to counteract the right-turning tendency of a descending airplane.

Lead the level-off by 50 feet for every 500 FPM indicated by the vertical speed indicator. Level off within 50 feet of the desired altitude.

LANDING

When landing maintain a pattern altitude within 50 feet of that specified for the airport in the *Airport Facility Directory*.

On base leg establish an approach speed within three knots of the manufacturer's recommended approach speed. This degree of precision requires correct trim, the pilot's awareness of pitch attitude, and a smooth reduction in power.

Select a landmark on a half-mile final and cross it within 50 feet of 500 AGL. This will prevent an approach that is either embarrassingly high or dangerously low.

Touch down on the second centerline stripe beyond the runway number. That stripe is 120 feet in length and provides a reasonable tolerance for accuracy.

Set your own tolerances for satisfactory performance in all segments of a flight. Each segment then becomes a practice drill to insure that your basic pilot skills will remain preflight-ready. Don't be afraid to set your standards high. It is just as easy to fly with precision as it is to fly without it. There is, after all, a limit to sloppy flying! And it is as easy, for example, to *maintain* a desired altitude within fifty feet as it is to *correct* two hundred feet to get back on the altitude. It is simply the pilot's individual choice.

A Monthly Training Flight

In addition to making every flight a practice flight, a pilot should conduct a once-a-month training flight. This training flight should be devoted entirely to the practice of a maneuver that has special significance in assuring preflight-ready pilot skills. Two such maneuvers are *turns about a point* and *stall recoveries*. Turns about a point help the pilot develop the ability to fly his plane accurately while his attention is distracted elsewhere. A distracted flier is, after all, the underlying cause for most pilot errors. A repeated and continuous stall recovery practice simply recognizes the fact that the stall remains the number one proximate cause of fatal aviation accidents.

These two maneuvers should be practiced in alternate monthly flights as a final step in an ongoing preflight program to help the pilot maintain and evaluate his basic pilot skills. An explanation of each maneuver follows to help direct your self-conducted recurrent training.

TURNS ABOUT A POINT

Turns about a point involve flying a circle around a landmark and correcting for wind drift so that a constant radius is maintained. Satisfactory performance means the turns are coordinated, altitude is maintained, and the plane rolls out of the maneuver on its original entry heading.

Turns about a point are designed to develop and test several skills that are particularly essential to low altitude (traffic pattern) operations. The pilot learns to fly an accurate ground track and maintain an exact altitude while his attention is occupied with matters away from the airplane. He also learns to maneuver accurately while his attention is confined inside the plane with an array of cockpit chores to accomplish. The pilot gains a new awareness of ground references; an awareness that enables him to see and use insignificant features on the ground that help him plan a flight path and stay oriented to it. The effort spent on this maneuver should teach the pilot how to quickly read the wind. He will learn to instantly detect a drift caused by a changing wind and how to correct for it. Wind is fickle. It rarely blows at a steady speed or from a constant direction. The pilot must learn to be flexible with his wind correction. He must be able to vary his corrections instantly to allow for the wind's whimsical nature.

The radius of the turn around the point should be between 700 and 1,000 feet. This is large enough to accommodate most light airplanes, yet small enough to necessitate a relatively steep bank. By the end of the flight the pilot should be able to maintain his altitude within fifty feet of that desired.

Basically, keeping a constant radius turn about the point depends on varying the rate of turn to match the varying ground speed. The faster ground speed requires a faster rate of turn to stay on the circular track.

The same thing happens in an automobile. Imagine an auto traveling thirty miles an hour. As it reaches a turn in the road, the driver only has to turn the steering wheel a slight amount. The auto's slow speed (ground speed) called for a slow rate of turn to stay on the road (circular track). In a car the rate of turn is determined by the amount that the steering wheel is turned.

Conversely, imagine the same car on the same road but now going ninety miles an hour. As the car reaches the turn a fast rate of turn is needed to match the fast ground speed. The driver must make a large swing with the steering wheel if he is to stay on the road.

The airplane is subject to the same rule concerning ground speed and rate of turn; the faster the ground speed, the greater the rate of turn required to stay on a circular course. With an airplane, however, the rate of turn is determined by the angle of bank. A greater angle of bank produces a greater rate of turn. The ground speed, of course, varies in the turn about a point as the plane moves from the tail wind position to the nose wind position. Thus, the airplane executing a turn about a point will require the greatest bank when it is in the direct tail wind position of the maneuver. The shallowest bank will be required when the plane meets the direct nose wind.

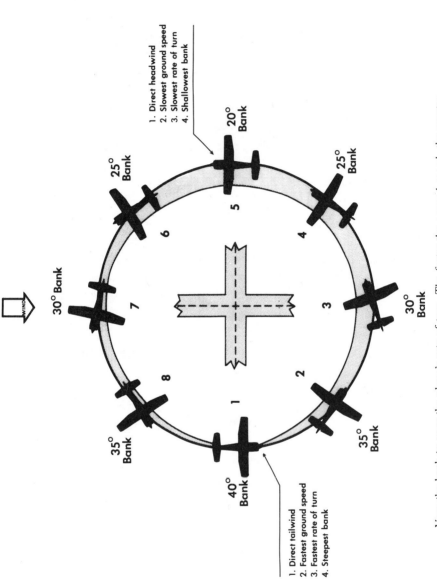

Vary the bank to vary the plane's rate of turn. The faster the ground speed, the greater the rate of turn required to stay on the circular course.

Keeping the Pilot Flight-Ready 85

An illustration will help describe how the pilot must vary his bank to match his ground speed.

For purposes of describing how the bank must be varied to accommodate a changing ground speed, let us start with the airplane at Point One in the illustration on page 84. (The banks indicated are for illustration only. It would take a definite combination of wind, radius of turn, and speed for these exact banks to accomplish a turn about a point.)

At Point One, the plane is in the direct tail wind position and will have its highest ground speed. Thus it is at this point that the steepest bank will be required (forty degrees in the diagram). As the plane passes through Point One, it immediately starts to lose its direct tail wind, and the plane's ground speed starts to decrease. This also means that as the plane passes through Point One, the bank must begin to shallow to match the diminishing ground speed. The ground speed will continue to diminish through Points Two, Three, and Four as the plane first loses the tail wind and then finally turns into the head wind. The bank must diminish as the ground speed diminishes.

At Point Five the plane is momentarily in a direct head wind and experiences its slowest ground speed. The bank is at twenty degrees, the shallowest at this point.

As the plane passes through Point Five, it immediately starts to lose its direct head wind, and the plane's ground speed starts to increase. This also means that as the plane passes through Point Five, the bank must begin to steepen, if it is to match the increasing ground speed. The ground speed will continue to increase through Points Six, Seven, and Eight as it loses the head wind and then starts gaining a tail wind. The bank must continue to steepen as the plane continues to gain ground speed.

Many pilots erroneously feel that the steep bank should occur at Point Three, and the shallow bank at Point Seven. The basis of this common error is the mistaken belief that the bank is varied in order to prevent wind drift. This, however, is not the case. The pilot prevents wind drift by crabbing, just as he did in tracking along a straight road.

Turns about a point give pilots more problems than any other primary maneuver of basic flying skills. Here are some of the most common errors along with some suggested remedies.

Common error no. 1: Many pilots will not picture the circular course on the ground. They get so involved with procedure that they forget about what they are trying to accomplish. They devote all their thought to wind direction, ailerons, and rudders and do not think about the circle on the ground that they are trying to track.

If the desired circular course were somehow painted on the ground, most pilots would have no trouble tracking it. Once they could *see* what they were supposed to be doing, they would no longer be concerned with wind direction or how and when to manipulate the controls. They would simply do whatever was necessary to stay on an easily recognized ground track.

It is essential therefore that the pilot visualize the desired circle around a center point. He must select and use landmarks that graphically lay out his desired course against the ground.

The easiest way to accomplish this depiction is by the use of a road intersection. The road intersection itself will serve as the center point. Four landmarks, one along each road, will serve as checkpoints. These checkpoints must be equidistant from the center point. It then takes little imagination to join the checkpoints together in a prominent circular ground track.

Common error no. 2: Many times a pilot loses mental track of the wind direction or becomes geographically disoriented after he has entered the maneuver. The road intersection helps him solve both problems. Before starting the maneuver, he can determine from ground references just how the wind direction intersects his crossroad. It will then be easy to keep tabs on the wind.

The crossroad also helps the pilot stay geographically oriented in the turn. He can use the roads for directional guidance. The pilot may easily enter the maneuver parellel to one of the roads, make the desired number of turns, and roll out parallel to the road on which he entered. This is an easy guarantee that his roll-out heading will match his entry heading. (A pilot should not make numerous 360s about the same point. Three turns about a single point are sufficient. Prolonging the maneuver any further may prove to be a discomfort.)

Common error no. 3: Some pilots let the altitude vary unreasonably while performing turns about a point. This usually happens for one of two reasons: either the pilot fails to realize that he must vary the stick pressure as he varies his bank, or he neglects his outside horizon reference during the maneuver.

The pilot must apply extra stick pressure as he steepens his bank in the maneuver. If he doesn't he will lose altitude. Similarly, he must relax stick pressure as he shallows his bank in order to prevent a climb.

It should be emphasized that both stick pressure and bank are constantly being changed during turns about a point. Some pilots try to think in terms of *steep bank* and *shallow bank* with no intermediate degrees of bank. The bank must constantly change simply because the wind and ground speed constantly change.

The pilot must use his outside horizon reference to maintain pitch control. The same sight-picture of nose against the horizon that he uses in his turns at altitude will serve him in his turns about a point. The pilot must learn to divide his attention between the ground reference points and the horizon reference. It takes a conscious effort on his part.

Common error no. 4: Some pilots seem to forget everything they learned about coordinating rudder and aileron, once they start the turns about a point. Their slips and skids usually develop for two reasons. First, they find that the wing tip covers the center point during part of the maneuver. They find themselves trying to move the wing tip with aileron while they maintain the circular track with forced rudder. Second, they sometimes get too involved with procedure, wind direction, and when and how to move the controls. They forget about the actual ground track that they are trying to accomplish. Both of these reasons for the slips and skids occur because the pilot has not adequately visualized the ground track by the use of four checkpoints equidistant from the center point. The selection of four suitable checkpoints equidistant from the center point is essential. This is because the wing tip will cover the center point during a portion of the turn. In addition, the center point will be hidden from the pilot's view during most of the time that the maneuver is being executed to the right. The four surrounding checkpoints will be needed for guidance during these times.

STALL RECOVERIES

The stall recovery is the second recurrent training maneuver in the ongoing program of preflight preparedness. A stall is a condition wherein the wing no longer provides sufficient lift to support the weight of the airplane. The airplane falls. Two situations can bring about this condition. First, insufficient airspeed, resulting in decreased lift, and second, an abrupt maneuver that increases the plane's aerodynamic weight (load factor) beyond the lifting capability of its wing. Several pilot errors can lead to either of these situations. Usually these errors lead to a combination of both low airspeed and high load factor. Some such errors are:

Neglecting pitch attitude when turning from downwind leg. A common mistake is neglecting pitch attitude when turning from a downwind leg. A pilot has several cockpit chores to perform when he is in the landing pattern. He must work communications, take care of flaps and landing gear, adjust power and prop, and reset trim. If he is negligent, the airplane may easily slip, unnoticed, into a stall attitude. This can happen when a pilot lets a high cockpit work load divert his eyes from where they should be: looking outside to keep tabs on his position, other aircraft, and his own aircraft attitude. Or it can happen at night when pitch attitude is a little harder to determine.

Executing a steep turn from base leg to final. Another error is executing a steep turn from base leg to final. This usually happens when the pilot has not taken a base tail wind into account. He starts drifting wide and feels that he must steepen his turn to prevent overshooting his final approach. If he isn't careful, he will compound the error by trying to reduce his bank with a skidding turn. He, of course, should have made a shallower turn, overshot final, and planned on landing a little further down the runway or made a go-around. Unfortunately, as long as we have airplanes and distracted pilots to fly them, this pilot error will continue to produce stall/spin accidents.

Trying to stretch a glide on final approach. This further error of trying to stretch a glide on final approach most often occurs when a pilot tries to clear low obstacles near the approach end of a short runway. He should, of course, add power. If a small application of power does not then correct the glide path, a go-around is called for.

A pilot should always execute a go-around rather than try to save a sloppy approach. It is safe to say that doing so could have prevented over 90 percent of the stall/spin accidents that have occurred in the landing phase.

Neglecting the pitch attitude when executing a go-around. Still another error is that of neglecting the pitch attitude when executing a go-around. On a go-around a pilot has no business other than maintaining airspeed, altitude, and directional control. This means holding the nose down level, applying full power, and staying on the right rudder. The go-around can hold hazards for the unwary pilot, but the main problem is pitch controls. When full power is applied, the nose tries to come up too high. If the pilot lets himself become distracted by a cockpit chore such as radio or flaps, he may not be aware of the changing aircraft attitude and a stall may develop.

There are many pilot errors that can lead to a stall. Very few of the stalls would actually occur if the pilot knew how to recognize that a stall was developing. Thus a pilot's main concern is that he be able to tell when a stall is approaching and prevent its full development.

The pilot should first practice recovering from partial stalls. The partial stall is signaled by mushy controls and the "buffet."

It often helps a pilot to recognize the buffet if he knows how it is produced. A buffet is produced when the smooth airflow over the wing is disturbed. The air then tumbles away from the wing. The buffet is a warning of an impending stall because the whole wing does not stall at once. The wing is designed so that the tips stall last. The stall starts at the wing root, next to the fuselage. As the airflow over the wing root is disturbed, it tumbles back and hits the horizontal tail plane that causes the buffet. (If a pilot listens carefully he may hear the eddies of wind striking the tail.) The buffet tells the pilot that the wing roots have stalled, and that the stall is progressing out toward the wing tips. One hopes he will never inadvertently let a stall develop beyond this point.

The practice flight should include work with stalls entered both with flaps and without flaps. Use a working altitude in

Keeping the Pilot Flight-Ready

excess of 3,000 AGL. Stalls in these configurations should be entered with power off and power on. No matter what the different configurations on stall entry, however, the threefold recoveries should be identical:

1. Pin the nose to the level attitude. It is not necessary to shove the nose down below level in a modern single aircraft. Remember, the habits learned at a safe altitude will be put into play on any stall that might develop close to the ground. A nose-down attitude could waste valuable altitude.

2. Apply all available power. Pilots are often surprised to find that by doing so they can recover from partial stalls with practically no loss of altitude.

3. Stop any roll that develops with coordinated use of rudder and ailerons. In older airplanes it was proper to stop the roll with only the rudder. The ailerons lost their effectiveness when the wing was near a stall and only complicated matters by creating extra drag at the wing tips. But it is a different story with modern airplanes. When a plane is certified, the manufacturer must demonstrate that the ailerons retain 10 percent effectiveness throughout a stall. Therefore, in a modern airplane use the ailerons in conjunction with the rudder to prevent the roll from developing. Doing so will speed stall recovery.

The pilot should work with partial stalls until he can recognize their onset and effect a smooth positive recovery. When the pilot is satisfied that he has demonstrated this level of ability, work can progress to full stalls. The same flap and power settings should be used. Recognition and recovery will remain unchanged.

In working with stall recognition and recoveries, it is important that the pilot strives to duplicate realistic conditions that lead to accidental stalls.

If, for example, the pilot has a takeoff stall in mind, he should picture how this type of stall is accidentally encountered. Most of these stalls happen on takeoffs from short runways with obstacles ahead. The pilot climbs out too steeply and continues to steepen his climb as the obstacle approaches. Then when he sees, under stress, that he is not going to clear the obstacle, he turns. The load factor produced by the turn, coupled with the low airspeed, produces the stall. Once the pilot pictures the sequence of errors, he should simulate similar errors. In doing so, he should create slow-flight at lift-off speed in order to simulate rotation from a short field. Then the pilot should apply climb power and begin a steadily steepening climb toward the imaginary obstacle. As the plane approaches the imaginary obstacle, he should simulate a turn away from it as he applies the remaining back travel of the stick. The stall that occurs will have been produced by a common series of pilot errors. The pilot must put forth an effort to make the stall exercises realistic demonstrations of actual common pilot errors.

For instance, if the pilot wants to work on recoveries from turning stalls, he should simulate what usually happens when a pilot, under stress, is turning from base to final and discovers in the turn that he is too low. That stall occurs when the pilot tries to hold altitude with back pressure. It is mandatory, then, when the pilot sets up this particular practice stall, that he apply flaps, approach power, and approach speed. Then as he starts a steep ninety-degree turn from a simulated base leg to final, he should try to prevent *any* loss of altitude with back pressure on the stick. This is the way it really happens. This presentation is far better than just going out and practicing accelerated stalls as they are described in training manuals. (In all cases, of course, review and comply with the manufacturer's handbook recommendations concerning stalls.)

Many pilots harbor an exaggerated mental picture of the stall attitude. To actually see the stall attitude for yourself, look at a parked tail-wheel airplane. Since these are landed full-stall, the parked plane rests in the true stall attitude.

Keeping Aeronautical Knowledge Preflight-Ready

Just as a pilot cannot expect to bring his pilot skill up to flight readiness in that hour before the flight, neither can he expect to refresh his aeronautical knowledge immediately before the flight begins. A continuous preflight program of updating knowledge is needed.

The pilot can decide between a formal classroom environment or a program of self-study. Either method is satisfactory, as long as it follows a logical schedule of learning. If you prefer the classroom environment, contact your area FAA General Aviation District Office to obtain the names of aviation ground schools in your locale. (FAA District Office phone numbers are listed in the *Airport/Facility Directory*.)

It is a good habit to try and attend a ground school every second or third year. Rules and procedures change, and we tend to forget those items of aeronautical knowledge if we do not put the information to work on a frequent basis.

If you prefer a program of self-study, consider using the study outline presented at the close of this chapter. It will program your study by knowledge areas as you read various texts and will serve as a checklist to insure that no items of pilot knowledge are overlooked.

Whether you decide on a program of classroom attendance or a schedule of home study, you must incorporate a means to constantly evaluate the preflight status of your aviation knowledge. To achieve this, purchase a copy of the FAA *Written Test Guide*. This guide presents test items in every area of aeronautical knowledge. There are several questions pertaining to each area, with each area covered on a single page. For an ongoing evaluation of your flying knowledge, give yourself a written exam every ninety days. On your first test, answer the first question on each page, and grade yourself with the answer key in the rear of the booklet. Then, on subsequent ninety-day tests, take the succeeding questions on each page in sequence.

As additional study material, every pilot should subscribe to the free Advisory Circulars offered by the FAA. These Advisory Circulars are designed to inform the pilot of current changes in procedures and also contain information that concerns aircraft safety. To include yourself on the AC mailing list, just send your name and request to:

> U.S. Department of Transportation
> Distribution Requirements Section, M–482–2
> Washington, DC 20590

In Review

- An ongoing program to keep the pilot flight-ready must perform two functions:
    1. Provide a schedule of updates, training, and study.
    2. Provide a means to evaluate the pilot's state of skill and knowledge.
- Make every flight a practice flight, with each segment of the flight serving as a practice drill:

    Decide on tolerances you are willing to accept as a measure of satisfactory performance.
    Reduce these tolerances to writing.
    Make your plane perform within the limits of your tolerances.

- Each pilot must determine for himself the accuracy he wishes to maintain in each segment of the flight:

    Takeoff
    Climb to altitude
    Level off to cruise altitude
    Cruise
    Descent from cruise altitude
    Landing

- It is as easy to fly with precision as it is without it; it is as easy to maintain a desired altitude within 50 feet as it is to correct a 200-foot error to get back on that altitude.
- In addition to making every flight a practice flight, a pilot should conduct a once-a-month training flight devoted entirely to the practice of a maneuver.
- Two maneuvers suitable for monthly practice flights are:

    Turns about a point
    Stall recovery

- Turns about a point help the pilot develop the ability to fly his plane accurately while his attention is distracted.
- Stall recovery practice simply recognizes the fact that the inadvertent stall at low altitude remains the number one cause of fatal accidents.
- Practice stall recoveries in excess of 3,000 AGL.
- Four common pilot errors that can result in a low-altitude inadvertent stall are:

    1. Neglecting pitch attitude when making traffice pattern turns
    2. Executing a steep turn from base leg to final leg
    3. Trying to stretch a glide on final approach
    4. Neglecting the pitch attitude when executing a go-around

- Just as a pilot cannot expect to bring his pilot skill up to flight readiness in that hour before the flight, neither can he expect to refresh his aeronautical knowledge immediately before the flight begins.
- An ongoing preflight program to maintain and update aeronautical knowledge may utilize either:

    Formal classroom lecture
    A program of self-study

- The FAA General Aviation District Office in your area can suggest ground schools in your locale.
- Use the questions in the FAA's *Written Test Guide* to evaluate the preflight readiness of your aeronautical knowledge.

PREFLIGHT AIDS

Aeronautical Knowledge Home-Study Required Subject Areas

I. Subject Area: Weather Reports and Forecasts
   A. Service Outlets
      1. Flight Service Station (FSS)
      2. Transcribed Weather Broadcast (TWEB)
      3. Pilot's Automatic Telephone Weather Answering Service (PATWAS)
      4. En Route Flight Advisory Service (Flight Watch)
   B. Surface Aviation Weather Report
      1. Sky condition and ceiling
      2. Prevailing visibility
      3. Weather and obstructions to visibility
      4. Sea level pressure reading
      5. Temperature and dew point
      6. Wind direction and velocity
      7. Altimeter setting
      8. Remarks section
   C. Pilot Reports
   D. Surface Analysis Weather Map
      1. Valid time
      2. Isobars
      3. High and low pressure systems
      4. Frontal symbols
      5. Station model symbols
   E. Weather Depiction Chart
      1. Sky coverage depiction
      2. Cloud heights
      3. Visibility

Keeping the Pilot Flight-Ready 99

    4. IFR areas
    5. MVFR areas
  F. Radar Summary Chart
    1. Echo pattern and coverage
    2. Associated weather
    3. Weather intensity and trend
    4. Movement of weather
    5. Heights of clouds
  G. Winds Aloft Charts
  H. Terminal Forecast
    1. Valid time of forecast
    2. Sky and ceiling
    3. Expected visibility
    4. Expected weather
    5. Expected wind direction and velocity
    6. Expected changes
  I. Area Forecasts
    1. Valid period of forecast and outlook
    2. Area of forecast
    3. Clouds and weather expected
    4. Icing conditions expected
    5. Outlook
  J. Sigmets and Airmets

II. Subject Area: Aviation Weather
  A. Nature of the Atmosphere
    1. Troposphere
    2. General circulation
    3. Uneven heating of earth
      a. radiation
      b. convection
    4. Temperature variations
      a. land/water

    b. day/night
    c. lapse rate
 5. Dew point
 6. Atmospheric pressure
    a. measurement
    b. isobars and pressure gradients
    c. lapse rate
 7. Pressure variations
    a. daily variations
    b. movement of pressure systems
    c. deepening of systems
 8. Hypoxia
    a. causes
    b. symptoms
 9. Air masses
    a. polar—characteristics
    b. tropical—characteristics
    c. continental—characteristics
    d. maritime—characteristics
10. Clouds
    a. cumulus (nimbus)
    b. stratus (nimbo)
    c. virga
    d. lenticular
11. Wind
    a. wind flow around a high pressure area
    b. wind flow around a low pressure area
    c. relationship to isobars and pressure gradient
    d. Coriolis force
B. Fronts
 1. Relationship to air mass
 2. Relationship to low pressure area

a. direction of movement
   b. trough
 3. Discontinuities across a front
   a. temperature changes
   b. atmospheric pressure drops
   c. wind shifts
   d. dew point spread lessens
 4. Significance of a front to a pilot
 5. Speed of frontal movement
   a. cold front
   b. warm front
 6. Normal cloud types
   a. cold front
   b. warm front
 7. Visibilities outside of precip
   a. cold front
   b. warm front
 8. Turbulence normally encountered
   a. cold front
   b. warm front
 9. Width of weather band
   a. cold front
   b. warm front
 10. Stationary front
 11. Occluded front
C. Thunderstorms
 1. Elements needed for formation
 2. Stages of a thunderstorm
   a. recognition of each stage
   b. significance of each stage
 3. Duration and size of thunderstorms
 4. Squall lines

D. Turbulence
      1. Over mountains
      2. Around obstacles
      3. Relationship to surface wind
      4. Around cumulus clouds
      5. Wing-tip vorticies
   E. Fog
      1. Radiation fog (ground fog)
         a. how created
         b. effect of wind
      2. Advection fog (sea fog)
         a. how created
         b. effect of wind
   F. Ice
      1. Rime ice
         a. elements needed for formation
         b. associated cloud types
         c. characteristics of rime ice
         d. factors affecting rate of formation
      2. Clear ice
         a. elements needed for formation
         b. associated cloud types
         c. characteristics of clear ice
         d. factor affecting rate of formation

III. Subject Area: Aircraft Instruments
   A. Static/Pitot System
      1. How static/Pitot system works
      2. Associated instruments
   B. Airspeed Indicator
      1. Significant speeds
         a. indicated airspeed

    b. calibrated airspeed
    c. true airspeed
  2. Instrument color coding
C. Vertical Speed Indicator
  1. As a trend instrument
  2. As a rate instrument
D. Altimeter
  1. Significant altitudes
    a. indicated altitude
    b. true altitude
    c. pressure altitude
    d. density altitude
  2. Atmospheric pressure
    a. altimeter setting
    b. effects of flying from one pressure area to another
E. Magnetic Compass
  1. Compass directions (the compass rose)
  2. Acceleration error
  3. Turning error
  4. Compass error
    a. deviation
    b. compass card
F. Altitude Indicator
  1. Operating limitations
  2. Caging
G. Heading Indicator
  1. Operating limitations
  2. Setting to the compass
H. Turn Indicator
  1. Standard rate turn
  2. Slip/skid indicator
  3. Relationship of speed and bank to rate of turn

IV. Subject Area: Radio Navigation
   A. VOR Navigation
      1. Ground equipment
         a. VOR radials
         b. line-of-sight transmission
         c. coded identification
      2. Airborne equipment
         a. frequency selector
         b. omni bearing selector
         c. to/from indicator
         d. left/right needle
      3. Obtaining a heading to the station
      4. Obtaining a heading from the station
      5. Intercepting an airway
   B. ADF Navigation
      1. Advantages over VOR
      2. Disadvantages of ADF
      3. Relative bearing to station
      4. Magnetic bearing to station
      5. Tracking to a station
      6. Tracking from a station
      7. Intercepting a desired bearing

V. Subject Area: Dead Reckoning Navigation
   A. Effect of Wind on Navigation
      1. Ground speed
      2. Drift
   B. Magnetic North vs. True North
   C. Isogonic Lines and Variation
   D. Compass Error and Deviation
   E. Measuring Course with Plotter
      1. Distance scale
      2. True course

F. Flight Computer
  1. Time—distance—ground speed
  2. Fuel consumption
  3. Wind correction angle
  4. Ground speed
  5. Conversion indexes

VI. Subject Area: Federal Aviation Regulations
  A. Selected FARs: Part 61
    FAR 61.15 Offenses involving drugs
    FAR 61.23 Duration of medical certificate
    FAR 61.31 General limitations
    FAR 61.51 Pilot logbook
    FAR 61.53 Operations during medical deficiency
    FAR 61.57 Recent flight experience
  B. Selected FARs: Part 91
    FAR 91.3 Responsibility and authority of pilot
    FAR 91.5 Preflight action
    FAR 91.11 Liquor and drugs
    FAR 91.14 Seat belts
    FAR 91.24 Transponders
    FAR 91.27/31 Aircraft documents
    FAR 91.32 Oxygen
    FAR 91.67 Right-of-way rules
    FAR 91.73 Aircraft lights
    FAR 91.79 Minimum safe altitudes
    FAR 91.81 Altimeter settings
    FAR 91.85 Operating on airport
    FAR 91.87 Operating in airport traffic area
    FAR 91.89 Operating at uncontrolled airports
    FAR 91.90 Terminal control areas
    FAR 91.105 Basic VFR weather
    FAR 91.107 Special VFR

FAR 91.109 VFR cruising altitudes
FAR 91.169 Inspections
C. Selected NTSB Part 830
830.5 Initial notification of accidents

VII. Subject Area: Radio Communications
   A. Radio Phraseology
      1. Aircraft identification
      2. Stating directions/altitude/speed
      3. Zulu time
   B. Communication Facilities
      1. Automatic Terminal Information Service
      2. Ground Control
      3. Tower
      4. Approach Control (departure)
      5. Flight Service Stations
      6. ATC Center
      7. Flight Watch
      8. Unicom
      9. Emergency
   C. Transponders
      1. Codes
      2. Squawk
      3. Ident

VIII. Subject Area: Airport Lighting and Marking
   A. Airport Lighting
      1. Rotating beacon
         a. daylight operation
         b. military airports
      2. Control tower light signals
   B. Airport Marking
      1. Traffic pattern indicator
      2. Closed airport

Keeping the Pilot Flight-Ready

      3. Wind direction indicators
   C. Runway Lighting
      1. Boundary lights—runway/taxiway
      2. Threshold lights
      3. Visual approach slope indicator
   D. Runway Marking
      1. Runway numbering system
      2. Closed runway
      3. Displaced threshold
      4. Centerline marking
      5. Unusable surface marking

IX. Subject Area: Aircraft Performance Charts
   A. Takeoff Charts
      1. Factors affecting takeoff distance
      2. Reading chart
   B. Cruise Charts
      1. Power settings
      2. Fuel flow
      3. True airspeed
      4. Endurance
   C. Landing Charts
      1. Factors affecting landing roll
      2. Reading chart

X. Subject Area: Weight and Balance
   A. Weight
      1. Licensed empty weight
      2. Payload
         a. weight of gas/oil
         b. placarded maximum area weights
      3. Gross weight
      4. Maximum allowable gross weight

B. Balance
      1. Weight of each load item
      2. Datum
      3. Arm
      4. Moment
      5. Center of gravity
      6. Center of gravity range
   C. Reading Weight and Balance Charts

XI. Subject Area: Theory of Flight
   A. Four Forces Acting on an Airplane
   B. Factors of Lift
      1. Shape of airfoil
      2. Velocity of airfoil
      3. Angle of attack
      4. Density of air
   C. Load Factors
      1. In a turn
      2. Effect on stall speed
   D. Drag
      1. Induced drag
      2. Parasite drag
   E. Thrust
      1. Torque
      2. P-factor
      3. Gyro force
      4. Slipstream effect
   F. Engines
      1. Fuel system
         a. proper grade
         b. contamination
         c. carburetor ice
         d. engine oil grades

2. Engine operations
   a. proper leaning
   b. power settings
   c. engine cooling
   d. malfunctions

XII. Subject Area: Publications
   A. *Airman's Information Manual*
   B. *Airport/Facility Directory*
   C. Advisory Circulars

XIII. Subject Area: Sectional Chart
   A. Aerodrome Information
   B. Radio Frequencies
   C. Airway Data
   D. Special Use Airspace
   E. Cultural Features
   F. Geographic Features
   G. Obstacles and Terrain Elevations

XIV. Subject Area: Safe Operation Practices
   A. Collision Avoidance
      1. Radar traffic advisories
      2. Proper pattern entry and exit
      3. Accurate position reporting
      4. How to see and be seen
   B. Wake Turbulence Avoidance
      1. Arriving
      2. Departing
   C. Filing Flight Plan
   D. Proper Use of Checklists

THE AIRPLANE

May God forgive you; the plane will not.

CHAPTER 5: Evaluating Aircraft Performance

YOU WOULD NEVER EXPECT to see an experienced pilot contemplate a departure from a short, hot runway, his nose to the prop of his plane and saying: "Listen engine, in a minute I'm going to lift off from this runway. Now remember, I've faithfully changed your oil every twenty-five hours and done everything else right by you . . . besides, there are some deserving people depending on this flight. So when I mash on that throttle, I want you to give beyond yourself. OK?"

An experienced pilot knows that he cannot depend on chance or special consideration from his plane. He knows he cannot expect his plane to make allowances and perform beyond its capability within a given situation, such as climbing safely away from a strip that is too short for the temperature, extending the range beyond the dictates of the fuel flow, or delivering an uneventful landing when loaded aft of limits.

A pilot's preflight planning must totally reckon the task asked of the plane against the facts and figures of the manufacturer's performance charts and tables. But yet the pilot must remember that the manufacturer's performance figures are only approximate. The variables affecting aircraft performance are often difficult to accurately standardize and evaluate; runway surface and the head wind component of a runway breeze are two such examples. The most nebulous variable of all is pilot technique.

This chapter discusses total preflight planning in terms of the airplane and the variable factors in four areas of aircraft performance: takeoff, cruise, landing, and weight and balance. Only by understanding how each variable affects performance can the pilot totally and professionally preplan by relating the performance table to real-life conditions.

Takeoff Performance

There are nine variables that affect takeoff performance:

Flap position
Runway surface
Aircraft weight
Head wind component
Field elevation
Temperature
Obstacles to clear
Abort/takeoff distance
Pilot technique

Flap position. The aircraft manual usually states the recommended flap setting for a normal takeoff. It may take careful reading, however, to find the recommended flap position to clear obstacles. But find it you must if there are trees to clear, because the flap setting that facilitates a shorter ground run often decreases climb performance.

Runway surface. Aircraft takeoff charts normally contemplate a dry, paved surface. But the times you are actually concerned with takeoff distance are usually when you are faced with a grass strip takeoff. If your aircraft manual does not provide grass-runway performance figures, use this rule of thumb: Add 20 percent to the paved runway distance for takeoffs from a dry grass surface; 50 percent if the grass needs mowing or the surface is wet. If the runway is wet and you expect freezing temperatures at altitude or destination, plan to leave your gear extended for a few moments after takeoff to air dry.

Another runway condition to consider is runway slope. Runway slope on a "one-way" strip is a difficult factor to pin down. The terrain surrounding nearly every airport is different because it almost never is a question of the *slope*. It is usually the ridge or mountainous upslope from the airport that stops all the action. Your best preflight action is a discussion with the airport operator; he knows how your plane will perform at his airport.

Gross weight. Remember the big surprise you got on your first solo takeoff? For many it was the alarmingly quick lift-off without the instructor's ballast aboard. The lighter the load your plane carries, the less time and distance it takes to accelerate to flying speed. Some aircraft manuals provide only maximum allowable gross weight performance figures. In this case apply a rule of thumb: Reduce the required distance by 5 percent for each hundred pounds below maximum allowable weight.

Head wind. An oncoming wind reduces the distance and time that it takes a plane to accelerate to flying speed simply because the wind flow already has the wing "moving" through the air before the plane begins rolling. Estimate your head wind component from the wind sock. Fifteen knots stiffens a sock; seven to eight has it drooping at a forty-five-degree angle. If the wind sock shows the breeze within thirty degrees of runway alignment, estimate the full velocity as your head wind component. If the wind lies between thirty and sixty degrees, count on half the velocity as a head wind. A wind that blows across the runway at sixty degrees or more has little head wind effect. If your aircraft manual does not provide for various head wind components, consider this rule of thumb: Reduce the zero-wind distance by 15 percent for each ten knots of head wind component.

Field elevation. The higher you are, the farther apart are the molecules, and the thinner the air. As a result the wing has to move faster to have the same mass of air flowing over it. At high altitudes, therefore, you will need greater distance for acceleration. But at the same time the engine has less air to breathe. This means less fuel to burn if you lean properly; if you do not lean properly an overrich mixture results. Either way it reduces the engine's output and acceleration suffers.

Light single-engine aircraft generally require an extra 25 percent of runway for each 1,000 feet above sea level.

Temperature. Heat produces the same detriment to takeoff performance as a high field elevation—thin air. The wings and engine just do not get a chance to do their best. As a general rule, add 10 percent to the required distance for each 25° F above the standard temperature at that elevation. (Estimate standard sea level temperature at 60° F; subtract 10° per 2,500 feet for standard at the field elevation.)

The greatest temperature problem comes, of course, when you combine a warm temperature with a high elevation airport. Most pilots are amazed at the decrease in performance the first time they take off from a high, hot runway. I consider any runway length less than 3,000 feet as a short field for light planes if the elevation exceeds 2,500 feet and the temperature is greater than eighty degrees. If you have not had the experience, make a hot-day takeoff from your home airport with the carburetor heat "on." This gives a good simulation of the performance you can expect at a field 2,000 feet higher.

Obstacle clearance. A pilot is nearly always faced with obstacles to clear anytime he is concerned with runway length. It is this meaningful value that should be extracted from the plane's takeoff chart as part of total preflight preparation. Remember, when using your table, that real obstacles off the end of real runways do not come in standard FAA fifty-foot heights. If you estimate the actual obstacle at twice, thrice, or quadruple the fifty-foot performance-table height, you must increase your clearance distance accordingly. To do this, take the difference between "ground run" and "obstacle clearance" and apply it to each fifty feet of the obstacle's true height.

If a pilot faces an obstacle from a short soft field, he faces a dangerous takeoff. The best bet is to wait until the surface dries out. But if a takeoff must be made from a runway that is both short and soft, use a defensive procedure that stops the takeoff before an accident occurs. For example, plan the total-to-clear-obstacle distance from your lift-off point. If the plane isn't off the ground by the time it reaches that point, simply shut down. To put this procedure into operation, a pilot must taxi down the runway to the departure end and estimate the distance from runway's end to the obstacle. Then the pilot must taxi back up the runway for the additional yardage needed to clear the obstacle. I recommend getting out of the plane and marking that point with a highly visible "flag" such as a weighted sheet of paper or your handkerchief. Then if your takeoff roll is still ground-bound at that point, stop the action.

Most light aircraft manuals employ the plane's best-angle climb speed over the obstacle. ("Best-angle" is the climb speed that produces the maximum altitude for the horizontal distance flown.) Some, however, recommend a higher speed, even at the cost of climb performance. The manuals state it this way because that plane's best-angle of climb speed is close to or below the plane's power-off stall speed. A power failure following takeoff at best-angle could be disastrous.

Abort/Stop distance. Before you begin your takeoff from any short strip, determine your stop distance and flag it with a marker along the runway. If your plane is not off by that point, you can abort the takeoff with little risk of damage. Run beyond that point, however, and you are committed. The plane *must* lift off or suffer damage. Very few light single-engine airplane manuals state an abort/stop distance. I suggest using the landing roll distance required for this value since lift-off speed and touchdown speed are nearly the same.

Pilot technique. If your proposed flight encompasses a short field and your short field experience is minimal or your skills rusty, the solution is clear: Get some short field training before you depart.

Cruise Performance

The next step in preflight preparation of the plane is to examine cruise performance. Good cruise performance depends on the pilot's understanding of engine power settings, fuel mixtures, fuel management, and proper control trimming. Power settings cannot be guessed at. In order for the engine to perform its best, the pilot must know how to interpret the manufacturer's engine charts. Improper fuel mixture (a subject often missing from student training) costs a pilot excessive engine wear and reduced range. The validity of the aircraft manual cruise charts rests on a properly trimmed aircraft; a plane flown sidewise just won't meet the expected performance stated in the tables.

Power settings. The printers who publish aircraft operations manuals each have a vice-president in charge of small print. It is his job to design the cruise-power-settings chart—unreadable in a bouncing cockpit to even the keenest eye and steadiest hand. It is wise, before you leave the ground, to print your own power charts on a card in big block letters and numbers. Write down the three power settings for the altitudes you intend to hold: 75 percent power for normal cruise; 65 percent for economy cruise; 55 percent power should you need to go for the distance. On the subject of power settings, let me dispel two old wives' tales about manifold pressure versus engine revolutions. First, there is an old saying that suggests a pilot run "square" settings for best engine performance. In other words, the saying suggests

Evaluating Aircraft Performance

we "match" manifold pressure with RPM such as 24 inches manifold pressure with 2,400 RPM. The second adage advises pilots never to exceed RPM square settings with the manifold pressure; 24 inches manifold pressure, for instance, calls for at least 2,400 RPM. Neither statement is true. Let the MP/RPM combinations listed in the aircraft's performance charts guide you. They have been flight tested and approved by both the engine and airframe manufacturers. A popular light plane with a 180 HP engine, for example, lists a cruise range of 2,100 RPM to 2,500 RPM with manifold pressures all the way from 18 inches to 24 inches—a 65 percent setting of 24 inches with 2,100 RPM at 4,000 feet as a specific. If the manufacturer recommends a setting, use it; they are the experts.

Occasionally there is a discrepancy between the engine manufacturer's table and the aircraft handbook. In this case, use the aircraft's figures. Engine installation often calls for a change in engine operation.

Pilots always find more than one MP/RPM combination that gives similar percentage of power at each altitude. Which to use? I suggest trying each one and selecting the combination that produces the least vibration in your particular plane; heat and vibration are the two mortal enemies of an aircraft engine.

Fuel mixture. Another consideration in preparing the airplane for flight is the fuel mixture. An aircraft engine produces its maximum power only if its fuel ratio or air to gasoline is approximately twelve to one (twelve pounds of air to one pound of gas). Two things occur in flight to disturb this balance. First, the air changes density as we change altitude. Second, the amount of gas flowing through the carburetor varies as we move the throttle to vary power. The pilot, then, must be prepared to correct his fuel mixture anytime he changes altitude or varies his power setting so that he maintains the proper mixture.

The figures in a cruise chart are accurate as far as percentage

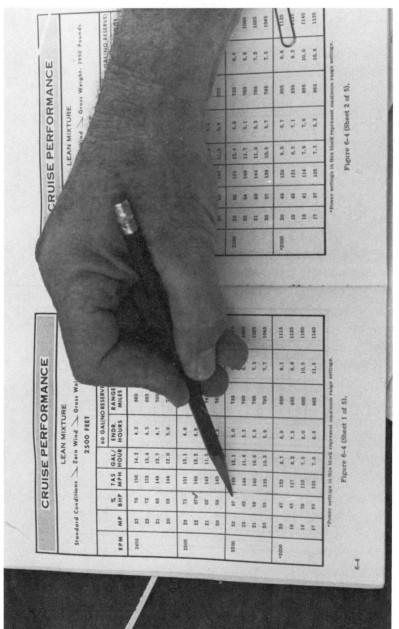

You always find more than one MP/RPM combination that gives a similar percentage of power at each altitude. Use the combination that produces the least vibration in your particular airplane.

of power, range, and endurance go only if the pilot maintains the correct fuel mixture. As an example, a popular 300 HP plane with its tanks topped off, cruising at 4,000 feet and 75 percent power, has an endurance of 5.1 hours with a properly leaned fuel mixture. This is the information that would appear in the cruise chart. The same plane, however, with a mixture control full rich can stay aloft only 4.2 hours. This bit of news could come as a rude surprise to a pilot who did not know the relationship that exists between fuel mixture and cruise chart and hence did not preplan on this happening.

There is an old and untrue tale which pilots hear that advises not to lean an engine's fuel mixture below 5,000 feet. This is false. The truth is, a mixture must be leaned at *any* altitude anytime the engine is producing no more than 75 percent power. Aircraft carburetors are factory-set to be slightly rich at sea level. Manufacturers purposely do this to protect the engine with the cooling effects of an overrich mixture during the brief moments of takeoff and climb-out when the engine produces nearly 100 percent of power. Any prolonged use of an overrich mixture, however, is tough on the engine; it runs rough and sets up excessive vibration.

To be totally prepared for leaning, a pilot should count on leaning the mixture for takeoffs and climbs only over 5,000 feet. At these higher altitudes the engine produces no more than 75 percent of its power; the thinner air combined with "full-rich" produces a mixture than even the most tolerant engine can't swallow.

Fuel management. Fuel management is another important area to be examined in the aircraft preflight checkup. Getting all the range and endurance promised by the cruise chart means getting all the gas from the tanks to the engine. Many "fuel starvation" accidents have cut short a flight with ample fuel aboard. A pilot's mismanagement of his supply can cause starvation by preventing fuel delivery from tank to carburetor. It is necessary for a pilot to study the fuel system section in his plane's manual. Even the pilot who flies the simplest of planes may find that the tanks should feed in a definite sequence, for example, or that takeoffs, climbs, or landings must be made from a certain tank with a minimum quantity.

The practice of running a tank dry before switching over is really not a safe operating procedure, though it is quite dramatic at that silent moment of switch. (The pilot moves briskly and efficiently, and the passengers usually snap to attention.) Pilots using the procedure run the risk of finding a vapor lock between the new tank and the engine, a trash-filled line, or a selector handle that either refuses to turn or falls off the moment they grab it.

Proper trimming. Unlike fuel management the proper trimming is not stated in the cruise charts; nevertheless the speeds and ranges in the cruise chart only come to a properly trimmed airplane. A pilot begins this proper trimming prior to takeoff. As every pilot knows, any power change calls for a retrimming. An out-of-trim airplane will fly sidewise through the air and induce a drag that the designer never contemplated. Trim the plane to fly straight as an arrow; fine tune elevator and rudder tabs so that the stick and rudder bars need only fingertip and toe-tip pressure to hold the plane's attitude.

Landing Performance

Many of the factors that influence takeoff performance affect landing in a similar manner and must be contemplated in your preflight work:

Flaps and airspeed. The manufacturer's estimate of landing distance is normally predicated on the use of full flaps and the recommended approach speed. If you choose to land with par-

Evaluating Aircraft Performance

tial flaps and a slightly higher approach speed, your preflight figuring needs to increase the landing chart distance by 10 percent for each five knots above the speed recommended on the chart.

If you use a speed in excess of "slightly higher," however, it is almost impossible to figure the distance required. You also run the risk of losing control of the plane by "wheelbarrowing" down the runway on the nosewheel, or swerving with uneven braking on a wet or rough surface.

Runway surface. The aircraft manual's landing distance is often stated for a paved braking surface only. As a preflight rule of thumb, add 20 percent to the landing distance for dry grass and 50 percent for wet grass.

Gross weight. If the manufacturer's charts do not provide for various aircraft weights, use a yardstick for modern single-engine aircraft while making preflight calculations: Reduce the required distance by 5 percent for each hundred pounds under the maximum allowable gross weight.

Head wind component. Each ten knots of head wind generally shortens the no-wind landing distance by 20 percent. Be sure to keep this in mind as you estimate your landing distance.

Field elevation. If the manufacturer does not specify various elevations in the aircraft's landing charts, increase the required sea level distance by 5 percent for each 1,000 feet of elevation in your preflight planning. Remember that high elevation landings may call for a leaned mixture on final to make a full-power go-around possible.

Temperature. Temperature and field elevation are not as critical to landings as they are to takeoffs; nevertheless, thin air must play into your preflight thinking. The "thin air" affects the wings' efficiency just as much on landings as on takeoffs, but on landings *engine output* is not a great factor. If you get the air hot enough the landing performance suffers. In the absence of manual information concerning hot days, use this general measurement: Add 10 percent of landing distance for each 40°F above the standard temperature for the elevation.

Obstacle clearance. Whether or not obstacles are actually present, the distance-to-clear obstacle figure is the one used to calculate the pilot's needed runway length for landing. If you think about it, the total-to-clear-obstacle distance only takes into account a fifty-foot obstacle, and surely any pilot wants to have at least fifty feet of altitude on short final. If you do have an obstacle to clear, remember to add a hedge factor if reason tells you that the obstacle is higher than the regulation-airport fifty-foot tree. To preflight the total distance required to clear higher obstacles, look at the landing chart and figure the difference between "obstacle clearance" and "ground roll." Then apply that difference for each additional fifty feet of obstacle height.

Abort/Go-around distance. Nearly every landing accident we hear of could have been prevented had the pilots made a go-around when things started going wrong. The reason those pilots did not make the needed go-around in simply because they didn't figure it into their preflight planning. Without preflight plans the pilots tried to make the critical decisions while skimming over rapidly evaporating runways, when suddenly the attempted landings quickly degenerated into a crash sequence, confusion and impact.

I recommend that one formulate go-around plans before starting any cross-country flight. Look at the landing chart and find the "landing roll" distance for the runway conditions you expect to encounter at your destination. Use twice that distance as your abort or go-around distance. (The extra yardage is needed to cover the possibility of a poor touchdown, excessive speed, or poor braking surface.) Then, while on the downwind leg, pick an object alongside the runway that marks the distance. (Knowing the runway length helps you visually estimate the needed distance.) If your plane is not in a perfect landing attitude with the wheels touching by the time you reach your abort point, simply make a go-around.

Weight and Balance

Several summers ago over south Florida, ATC received a call from a training flight. The instructor advised that he and the student (a commercial pilot going for his CFI) were going to commence a practice spin. They were to start from 4,200 feet and planned to have the recovery completed by 3,000 feet. The plane was a popular four-place trainer-type of 150 horsepower.

A few minutes after the initial call, ATC received this one: "Mayday! Mayday! Mayday! It won't come out!" And it didn't.

Months later, a similar accident in a similar low-performance trainer happened. Here an extremely experienced instructor had conducted stall recovery practice with his primary student. The plane stalled, spun, and never recovered.

The investigations of both accidents turned up some facts that make a pilot stop and think. First, each aircraft was a type that we often call "forgiving." Second, depending on the weight and balance, each plane could either be a normal category or a utility category aircraft, with spins approved only if loaded in the utility balance range. Third, each cockpit had a capable pilot at the controls. Fourth, while each aircraft was loaded within the normal category limits, it had exceeded the aft limits for utility. And fifth, each plane exceeded these aft limits by less than one-half inch. Interesting, isn't it, just how critical weight and balance can be. And yet departing pilots are seldom seen with their noses buried in the weight and balance charts doing their preflight calculations as the suitcases and passengers climb aboard. Of course, grief does not come to all planes that are slightly out of balance. It usually takes the onset of a stall to trigger the disaster, such as one produced by the load factors combined with the slow speed of a plane S-turning on final for spacing. Other examples are a pitch attitude that goes unnoticed during an emergency go-around or the arriving IFR flight with just a tad of ice on the wings. To guard against the disastrous combination of improper balance and an inadvertent stall, learn how to use your plane's weight and balance data. If you are unsure about the procedure, have an instructor guide you through some sample loadings.

When should a pilot calculate weight and balance? As a rule of thumb, during the preflight checkup run a weight and balance calculation anytime the seating capacity is over half filled, or when there is any luggage in the baggage compartment.

And, of course, run a careful calculation before any training flight that involves stalls, no matter how "forgiving" the plane, or how the weight aboard is distributed. The plane that I fly day in and day out is a garden variety two-seat trainer—about as docile as they come. Yet with a student of my equal weight aboard (I'm ashamed to say how much) and the seats shoved back to accommodate my big feet, the plane is unsafe to stall. It is just barely aft of limits—a little less than half an inch, as a matter of fact.

For the sake of his passengers, himself, and his plane, a pilot cannot afford to guess at his plane's performance. To do so is to rely on chance. And the very nature of chance is that it must someday turn against us. The preflight research that is necessary to evaluate the plane's capability against the task we set before it takes a certain amount of time and effort, certainly. But to fly with safety, there is no other way.

In Review

- A pilot's total preflight planning must gauge the task against the facts and figures of the manufacturer's performance charts and tables.
- Nine variables affect take-off performance:

    1. Flap position
    2. Runway surface
    3. Aircraft weight
    4. Head wind component
    5. Field elevation
    6. Temperature
    7. Obstacles to clear
    8. Abort/takeoff distance
    9. Pilot technique

- When obstacles stand ahead, remember that extended flaps often decrease climb performance.
- Grass runways normally add 20 percent to the paved-runway distance needed for takeoff; 50 percent if the grass needs mowing or is wet.
- As a rule of thumb, reduce the required takeoff distance by 5 percent for each hundred pounds below maximum allowable gross weight.
- Estimate your head wind component from the wind sock. Fifteen knots stiffens the sock; seven to eight knots hangs it at a forty-five degree angle.
- If the wind is within thirty degrees of the runway, estimate the full velocity as your head wind component; thirty to sixty degrees, half the velocity; sixty to ninety degrees, no effective head wind.
- As a rule of thumb, reduce the zero-wind takeoff distance by 15 percent for each ten knots of head wind.

- Single-engine aircraft generally require an extra 25 percent of takeoff distance for each 1,000 feet above sea level.
- As a general rule, add 10 percent to the required takeoff distance for each 25°F above standard temperature.
- Consider any runway length less than 3,000 feet as a short runway for light singles if the elevation exceeds 2,500 feet or the temperature is greater than eighty degrees.
- Remember that real obstacles do not come in standard fifty-foot heights.
- Use the "landing roll" distance for the abort/stop distance required.
- If your proposed flight encompasses a short field and your short field skill is rusty, get some short field training before you depart.
- Good cruise performance depends on the pilot's understanding of:

   Engine power settings
   Fuel mixtures
   Fuel management
   Proper control trimming

- Use only factory recommended power settings.
- Cruise charts take into consideration a leaned fuel mixture.
- The fuel mixture must be leaned at any altitude anytime the engine is producing no less than 75 percent power.
- A pilot should lean even for takeoffs and climbs if they occur over 5,000 feet.
- The practice of running a tank dry in the air is not a safe fuel management procedure.
- The speeds and range of the cruise chart come only to a properly trimmed aircraft.
- When using slightly higher than the factory recommended ap-

Evaluating Aircraft Performance

proach speed, count on an extra 10 percent landing distance for each extra five knots.
- As a rule of thumb, add 20 percent to the landing distance for dry grass; 50 percent if the grass is wet.
- In general, if the chart shows only maximum gross weight landing distance, reduce the distance 5 percent for each 1,000 pounds under maximum allowable weight.
- A general measure concerning elevation: Increase the sea level landing distance by 5 percent for each 1,000 feet of field elevation.
- Concerning temperature, add 10 percent to the landing distance for each forty degrees above standard temperature.
- Whether or not obstacles are actually present, it is the total-to-clear obstacle figure that provides the needed runway length.
- Almost all landing accidents that occur could have been prevented if the pilot had made a go-around when things started going wrong.
- Use twice the "landing roll" distance as your "go-around" distance. Pick a mark on the runway to establish this go-around point while you are on the downwind leg.
- If your plane is not in a perfect landing attitude with the wheels touching by the time you reach your "go-around" mark, then abort your landing.
- Another rule of thumb: Run a weight and balance calculation whenever the seating capacity of your plane is over half filled or there is any luggage going into the baggage compartment.
- Run a careful weight and balance calculation before any training flight that involves stalls.
- For the sake of his passengers and himself, a pilot cannot afford to guess at his plane's performance; to do so is to rely on chance.

Preflight Aids

TAKE-OFF PERFORMANCE

AIRCRAFT _____

FLAP POSITION _____

CLIMB SPEED 10 OBSTACLE ____

DISTANCES TO CLEAR A 50 FOOT OBSTACLE

At sea level:

GROSS WEIGHT	HEAD WIND	PAVED RUNWAY 59°F.	85°F.	Abort	GRASS RUNWAY 59°F.	85°F.	Abort
MAX. ALLOW.	0						
	10 Kt.						
	20 Kt.						
OTHER:	0						
	10 Kt.						
	20 Kt.						

At 3000 feet:

GROSS WEIGHT	HEAD WIND	PAVED RUNWAY 48°F.	85°F.	Abort	GRASS RUNWAY 48°F.	85°F.	Abort
MAX ALLOW.	0						
	10 Kt.						
	20 Kt.						
OTHER:	0						
	10 Kt.						
	20 Kt.						

At 6000 feet: (Fuel flow _____)

GROSS WEIGHT	HEAD WIND	PAVED RUNWAY 38°F.	85°F.	Abort	GRASS RUNWAY 38°F.	85°F.	Abort
MAX. ALLOW.	0						
	10 Kt.						
	20 Kt.						
OTHER:	0						
	10 Kt.						
	20 Kt.						

Calculate the takeoff distance your plane requires under various situations. This will give you a quick preflight estimate of the distance needed.

CRUISE PERFORMANCE

AIRCRAFT _____

55% Power

ALTITUDE	MP	RPM	FUEL FLOW	TAS	FULL TANK ENDURANCE	PARTIAL TANK ENDURANCE
2500					:	:
5000					:	:
7500					:	:
OTHER:					:	:
				DESIRED ENDURANCE RESERVE AT DESTINATION		:

65% Power

ALTITUDE	MP	RPM	FUEL FLOW	TAS	FULL TANK ENDURANCE	PARTIAL TANK ENDURANCE
2500					:	:
5000					:	:
7500					:	:
OTHER:					:	:
				DESIRED ENDURANCE RESERVE AT DESTINATION		:

75% Power

ALTITUDE	MP	RPM	FUEL FLOW	TAS	FULL TANK ENDURANCE	PARTIAL TANK ENDURANCE
2500					:	:
5000					:	:
7500					:	:
OTHER:					:	:
				DESIRED ENDURANCE RESERVE AT DESTINATION		:

Jot down the cruise power settings for easy in-flight reference.

LANDING DISTANCE

AIRCRAFT _____
APPROACH SPEED OVER OBSTACLE _____
FLAP POSITION _____

DISTANCES TO LAND OVER A 50 FOOT OBSTACLE

At sea level:

GROSS WEIGHT	HEAD WIND	PAVED RUNWAY 59°F.	85°F.	Abort	GRASS RUNWAY 59°F.	85°F.	Abort
MAX. ALLOW.	0						
	10 Kt.						
	20 Kt.						
OTHER:	0						
	10 Kt.						
	20 Kt.						

At 3000 feet:

GROSS WEIGHT	HEAD WIND	PAVED RUNWAY 48°F.	85°F.	Abort	GRASS RUNWAY 48°F.	85°F.	Abort
MAX. ALLOW.	0						
	10 Kt.						
	20 Kt.						
OTHER:	0						
	10 Kt.						
	20 Kt.						

At 6000 feet: (Fuel Flow _____)

GROSS WEIGHT	HEAD WIND	PAVED RUNWAY 38°F.	85°F.	Abort	GRASS RUNWAY 38°F.	85°F.	Abort
MAX. ALLOW.	0						
	10 Kt.						
	20 Kt.						
OTHER:	0						
	10 Kt.						
	20 Kt.						

Calculate the landing distance required under various representative situations and use the tables for quick in-flight estimates.

SAMPLE LOADING SITUATIONS
BAGGAGE · PASSENGERS · FUEL
AIRCRAFT : N_____

A. BAGGAGE WEIGHT ALLOWABLE WITH :

1. All seats occupied / Full tanks. _____
2. Front seats occupied / Full tanks. _____
3. All seats occupied / Partial (____ gal.) tanks. _____
4. Front seats occupied / Partial (____ gal.) tanks. _____

B. REAR-SEATS WEIGHT ALLOWABLE WITH FRONT SEATS OCCUPIED AND :

1. Baggage to capacity / Tanks full. _____
2. Baggage to capacity / Partial (____ gal.) tanks.
3. No baggage / Tanks full. _____
4. No baggage / Partial (____ gal.) tanks.

C. FUEL WEIGHT ALLOWABLE WITH :

1. All seats occupied / Baggage to capacity. _____
2. All seats occupied / No baggage. _____
3. Front seats occupied / Baggage to capacity. _____
4. Front seats occupied / No baggage. _____

Work several sample aircraft loadings and retain the results for a quick preflight reference.

CHAPTER 6: Checking the Plane for Flight

SEVERAL YEARS AGO, at a rural New England airport, a group of transient pilots were gathered in the fixed base operator's lobby, looking out at the flight line. The object of their attention—and amusement—was the lanky, elderly pilot meticulously preflighting his small two-seater, checklist in hand. His preflight was not only meticulous, but it might also have been called embarrassingly fastidious. He had been worrying over the little ship for nearly fifteen minutes. The onlookers enjoyed every moment of it; the way he would closely examine a detail, step back for an overall look, then rush forward for a nose-to-fabric examination of that detail as if trying to catch a discrepancy by surprise. Not a safety wire, nor an inspection plate, nor a spot of oil escaped his scrutiny.

The spectators were particularly entertained when the old pilot had to bend his spare frame double to peer up into one of the plane's inaccessible hidey holes.

A gray haired A&E trudged through from the shop area, carrying a tire out to the flight line. One of the pilots pointed out the window and called to the mechanic, "Who is that old worrywart?"

"Oh," replied the mechanic, "that's Charles Lindbergh." And trudged out the door.

The proper and professional preflighting of an airplane, as illustrated by this little anecdote, depends upon three elements:

1. willingness to expend the necessary time and effort
2. use an adequate written checklist
3. positive *attitude* toward preflight preparations

First of all, do not hurry any phase of your preflight. Get to the airport well ahead of departure time so you feel no need to rush through the checklists. The effort needed to conduct a proper preflight rarely requires physical exertion. Rather, it is an effort of concentration. When you conduct any phase of the aircraft preflight, direct your entire attention to your plane and its checklist. Do not talk to passengers, worry about upcoming navigation, or let any other distraction rob your preflight inspections of efficiency. Make sure your check is methodical. If you find that you are missing some items, or checking some out of sequence, be warned that your mind is wandering; stop your inspection at that point and start again at the top of your written checklist.

A pilot's attitude toward preflighting an aircraft is critical to the effectiveness of his inspection. Often a pilot conducts his preflight with the preconceived notion that the plane is probably OK. The pilot who shows up fifteen minutes before the flight to inspect his plane is committing an error. This pilot obviously does not expect to find anything wrong that requires service or repair. An inspection made with this frame of mind is only of marginal value.

In truth, almost every plane on almost every flight has a discrepancy that may breathe tragedy into the flight. Conduct your checks with this in mind. Be suspicious and let stubbornness work you through the checklists to ferret out the hidden hazards.

Aircraft Preflight Inspection

Many pilots conduct their aircraft's preflight inspection without being too sure which items they need to check or what clues point out a discrepancy. In most cases this happens simply because *preflight* was not adequately taught during the pilot's student training. All too often the student pilot was left on his own to develop and practice his preflight routine. Except for the first lesson or two (when the student was still unable to fully understand), the instructor was often not in attendance during the student's walk-around inspection. And as a result many students never learned *which* items to check and *why*.

A rated pilot cannot continue to rely on a haphazard preflight inspection to see him safely through his flying career. This leaves safety purely to chance, and the very nature of chance dictates that it must sooner or later turn against us.

As you formulate your own preflight routine, let's look at all the checklist items—why the items are critical to safety and how to perform effective inspections of them.

BEFORE THE INSPECTION BEGINS

Switches off and remove key. An ignition switch in an "on" position can turn the upcoming inspection of the propeller into disaster, as the engine could crank. You cannot tell at a glance if the key is "on" or "off" of one of the mags. Fortunately, today's modern switches have a built-in safety feature; you cannot remove the key unless it is in the "off" position. So—don't insert the key until you have yelled "clear" and are ready to start the engine. With that in mind, develop a habit after shutting down: Remove the key before you unfasten the seat belt.

Remove the gust lock. Before you start the walk-around, free the control surfaces. Otherwise, it is nearly impossible to inspect the aileron and elevator attachment points.

Set the parking brake. A light breeze (and certainly the prop blast from a nearby plane) can start a light aircraft rolling—

You cannot tell at a glance if the switch is "off" or "on" one of the mags. Do not preflight the plane with the key in the switch.

possibly right into a parked airplane. Always set the parking brake. Even a gentle tap brings expensive repairs.

Gear handle down and locked. If a gear handle or switch is in the "up" position, it is very possible to pull the wheels out from under the plane during the starting sequence. It's true that a safety switch on the retract mechanism is *supposed* to prevent the gear from retracting when the plane's weight rests on it. But these switches are normally checked only once a year, and they can go bad between annuals. Make sure gear handle is down and locked.

Remove the tie-downs or chocks. Untie all the ropes before you get involved with the preflight inspection. A pilot who detaches each line separately as he reaches that position stands a reasonably good chance of overlooking a tiedown rope. Taxiing with one wing still tied will swing you into any adjacently parked plane before your reaction time can prevent it. You cannot imagine how fast it happens. If it ever does, just look toward the hangar—because here he comes! Eight foot four, 380 pounds, naked, eating a raw chicken—and you just smashed his little airplane.

WING SECTIONS

Ailerons. Move the control surfaces by hand and feel for any binding. Ask a mechanic which attachment points must be safety-wired or pinned.

Fuel sumps. I have a commercial pilot friend—a high-time captain with all the caution of twenty-seven years' experience—who commutes some fifty miles each day by light plane to his airline's departure airport. One spring morning he discovered on his normal preflight inspection that his four-placer had not been refueled the night before. Suspecting condensation in the partially filled tanks, he drained the sumps with extra care. The preflight inspection was complete; he ordered the plane refueled.

The taxi and runup were uneventful. But in that moment following lift-off, a slug of water reached the engine and it quit. My commercial pilot friend only had time for a quick appraisal of the runway length remaining . . . time only for back pressure on the control wheel, the "thud" of a dead-stick touchdown came along with hard braking just before the plane ran off the end of the runway into the rough. No one was hurt and the gear received only minor damage—but where had the water come from?

Most pilots realize that it is necessary to drain the sumps on a plane that has been tied down overnight. They are aware of the condensation that can form in partially filled fuel tanks, and they can picture the sudden silence that this water creates on takeoff. Not all pilots are aware that those big tanks on the refueling trucks also develop condensation. If the lineman forgets to drain his refueling tank, he can easily pump water into your tanks. For this reason, drain the fuel sump *after* each refueling even though the plane was preflighted or flown just prior to topping off the tanks.

The Pitot tube. Bugs like to build nests in Pitot tubes, and a plugged tube renders the airspeed indicator unreliable. Fish out any insect deposits with a short piece of fine wire.

The stall-warning indicator. When preflighting some airplanes, you only need to turn the master switch on, trip the stall warning vane on the wing, and listen for a horn. On other models you need to prop the vane open with a small wad of paper while you look inside the cockpit for a warning light. (Use a ten dollar bill for your paper wad, and you won't forget to remove it before you fly.)

Fuel tanks. Looking down into a tank is the only way to

make sure there is fuel in it. Fuel gauges do malfunction, and many gauges register "full" even though an hour's gasoline had been burned from the tank. If your plane's gas tank does not have a "partial quantity" gauge or tab in the tank's filler neck, make yourself a dipstick. You can do this by dipping a ruler into the tank when you have burned a known quantity from the tank and then notching the stick.

Look for the proper octane color coding and smell for the kerosene odor that warns of a tank erroneously filled with jet fuel.

Wing tips. If any part of an airplane is apt to get damaged on the ground, it is usually the wing tip. Uncautious cars and refueling trucks, taxiing aircraft, even pedestrians seem attracted to the wing tips. Further, as fiberglass wing-tip fairings age, they often require stop-drilling to keep the crack from spreading.

LANDING GEAR

Tire inflation and wear. Don't just pump up a flat tire and go flying. If a tire is flat it has a leak, and pumping it tight only assures you of an uneventful takeoff. After the air escapes during your flight, you may make a landing that is discussed for weeks around the hangar.

Tread wear should be checked when inspecting the tires. The tread on aircraft tires serves a different purpose from tread on automobile tires. Tread on an auto tire is important for traction, and it must be deep for safety. On an aircraft tire, however, the tread is there primarily to show the depth of rubber remaining. Aircraft tires are manufactured on the thin side for lightness. Anytime you notice a tread line that has been completely worn from sight, count on the next layer being air and don't fly the plane until the tire is changed.

Checking the Plane for Flight 141

Brakes. During your walk-around inspection, look for the dirty residue of hydraulic fluid around the inside of the main wheels which will indicate a leaking wheel cylinder. Further, make a habit to check the brakes for evenness and strength as you taxi from the tie-down; simply pull ahead three feet and stop the plane before you continue taxiing.

Wheel struts, wheel-well doors, and the retract mechanism. Struts that are not properly inflated place undue wear and tear on the plane, and can make ground handling difficult. Normal inflation for many light aircraft allows you to move the strut up and down, yet your full weight should not completely compress it. Check the wheel-well doors for tufts of grass or any other foreign matter that could jam them. Under freezing conditions you should be concerned with water. After takeoff air-dry the wheel wells before you retract the gear. The same problem can arise when you take off with a wet gear from a warm airport and quickly climb to the freezing level, continuing to a destination that is below freezing. In this case it is possible to have trouble extending the landing gear.

If you do fly a retractable airplane, you may want to watch a retraction test at the maintenance shop on a model similar to yours. This demonstration will let you see exactly how the system works. A few suggestions from the mechanic will add understanding to the preflight inspection.

NOSE SECTION

Windscreen. Make sure you have a clean windscreen. Trying to look for traffic against a low sun through a dust-covered windscreen is an exercise in frustration. Your only wish will be that you had spent time cleaning it before the flight. Fly specks and bug smashes provide their own form of entertainment—if you catch them with the corner of your eye and try to dodge them! (A sure sign that it's time to land is when you distinguish between single-engine and multi-engine fly specks!)

The best way to clean a windscreen is with the palm of your hand and plenty of water poured from a bottle. Avoid any vigorous rubbing, because the accumulated grit can mar the soft Plexiglas. It goes without saying, don't try to wipe off the dirt with a dry rag. Let the surface air-dry. If you polish a windscreen dry, a static charge builds up that attracts dust and dirt.

If you are flying at night, clean the inside film from the Plexiglas as body humidity and the film of grease can turn the cockpit IFR.

Cowl flaps. Check to see that the control rod and attachment points are not excessively worn. Visit a maintenance shop and have the mechanic show you a cowl flap that is worn beyond safe service.

Propeller and spinner. A ragged nick in the propeller can create a fatigue line in the blade. You can relieve this metal stress by filing the nick smooth. Check the spinner and make sure that any small cracks have been stop-drilled to prevent losing it at 2,400 RPM.

If the plane has not flown for several days, of if the day is cold, pull the prop through each compression stroke to help lubricate the cylinder walls. This alleviates the drain on the battery when you crank up and minimizes cylinder wear in those first few critical seconds of engine operation.

Inductive air intake. Any mud or grass covering the inductive air intake can restrict full takeoff power. Be sure the intake is clean or you will soon discover your error on lift-off from a short runway.

Engine cooling inlets. Birds do build nests in the engine cooling inlets, and mechanics occasionally leave a rag resting inside. Either bird nests or rags easily catch fire in flight, and they will send back a whiff of smoke that ages the pilot prematurely!

Proper engine temperature depends on the correct baffling. A trip to see an uncowled engine in the maintenance shop and a few words with the mechanic will fill you in on how the engine baffles are arranged for proper power-plant cooling.

Cowling interior and exterior. Aside from checking the oil level, many pilots are not sure just why they are peering down into the cowling. Your inspection under the cowl should look for the pool of oil that warns of a leak, spots of fuel dye that mean gasoline is dripping where it shouldn't, or the white-line residue that signals a cracked exhaust manifold. If any of these symptons show up, be sure to consult the mechanic before the flight.

I find it valuable to use a rule of thumb concerning the oil level in most small engines: Use a minimum operable oil level of two quarts below maximum. As a result, an engine holding nine quarts maximum can operate with seven quarts in the sump—for a short flight. Remember, though, that many engines easily consume a quart in five hours of flying. So if your flight is a long one, bring the oil right up to full. An old canard in flying circles warns not to run the engine full of oil because it will "throw" the last quart. This is simply not true. An engine depends on adequate oil for internal cooling as well as for lubrication. You will find on a hot day that a final quart of oil on the dipstick lowers the cylinder-head temperature gauge by several degrees.

In a preflight checkup, when you drain the fuel strainer, be sure to draw out enough gasoline to catch any water that may lie in the lines between the tank and the engine. I have often drawn a pint before the water showed up, as these fuel lines are longer than you might suspect. If your plane has selective tanks, sample both lines by moving the selector valve.

In addition, during your cowling inspection evaluate the condition of all visible wiring, belts, and hoses. If any of these items shows appreciable wear, ask a mechanic's opinion about replacement before you fly the plane. Look, also, to see that all outside cowl fasteners are secure and that the cowl screws are in place to make sure the cowl does not open in the air.

EMPENNAGE

Control surfaces. In the preflight checkup be sure to move the elevator and rudder by hand, feel for any binding, and check the attachment points for proper safety wiring and pinning. Any frayed control cable is cause for canceling the flight.

Drain ports. Planes can collect a bucket or so of water when driving rain forces its way into the fuselage via the openings around the rear control surfaces. This water must find its way out of the airframe before flight; consider what a gallon of water in the tail does to the moment, or the possibilities that freezing temperatures offer.

Most manufacturers drill small drain ports in the bottom of the fuselage near the tail section, and often in the bottom of the rudder fin itself. To properly do their job, these small ports must remain open. The dirt and oil that collect on the plane's belly, however, often clog them. Take a sharp pencil point and poke the ports to make sure that they are open. If a stream of water flows, press the tail section down slightly to aid complete drainage.

FUSELAGE

Static pressure port. The airspeed indicator, altimeter, and vertical speed indicator all register from static pressure, and their correct readings depend upon an unobstructed port. Clear the small port of wax residue or dirt with a pencil point or match stick.

Tow bar. A tow bar lying loose in the baggage compartment makes an effective wrecking bar when the turbulence turns from moderate to severe. If left on the cockpit's hat shelf, a tow bar could injure a passenger. Make sure the tool is properly stowed in its rightful place.

Emergency locator transmitter. Among your preflight tasks, check to see that the ELT is in the "armed" position that triggers the transmitter on impact. Regulations allow you to test the ELT during the first five minutes of each hour. To do this, simply turn your aircraft radio on, set the frequency on 121.5, and turn the ELT to "on" or "test." If your ELT is working properly, the audio tone you hear is unmistakable. Procedure allows for a test of only three audio sweeps.

Baggage door. A baggage door that pops open in flight usually causes the rear-seat passenger to opt Greyhound for the return trip. So to help maintain general aviation's good name, be sure the hatch is secure and locked!

Exterior and interior lights. When a daytime flight extends to a nighttime flight, one is reminded why preflighting the lights is essential. It is easy to forget to preflight the lights. It is a further reason that a written checklist is important.

While on the subject of lights, make sure that you have spare fuses aboard. At checkup time run through a daylight preflight rehearsal of replacing blown fuses. Remove and replace each fuse for the drill. You will find that this simple task may not be so simple when you need to do it aloft in a blacked out cockpit with a flashlight clenched between your teeth. The preflight rehearsal, you will see, makes the in-flight task easier.

Another suggestion in the light department: If you plan a nighttime departure, consider carrying out aircraft preflight inspection earlier in the day. It is easy to let a discrepancy get by you in the dark. (On those occasions that a nighttime preflight is forced upon me, I am almost certain that I can hear the plane chuckle as I taxi away from the ramp.)

Fire extinguisher. A fire extinguisher is cheap insurance against that most dreaded pilot fear that has remained with aviators since the dawn of powered flight—in-flight fire. Just make sure that the extinguisher is secured in a sensible manner and will not become a cannonball when turbulence starts lifting things in the cockpit. Make certain, too, that the extinguisher is charged and approved for in-flight use. Many of the household extinguishers displace the oxygen in the cockpit and put out the pilot along with the fire. Others contain a propellant that creates instant IFR the second the trigger is pulled.

Overall appearance of the airplane. Pilots have been known to attempt takeoffs with cowlings or rudders missing. Step back and take an overall look at your plane from several angles. You can, for instance, inspect the upper wing surface of a high-winged airplane by stepping several paces behind the tail. Look for obvious defects. Is anything missing?

Starting the Engine

Competent preflight checks and inspections do not depend on a pilot's familiarity with the aircraft; they depend upon his familiarity with his own checklists. To find out how true this is, take your written lists to an unfamiliar aircraft model and put them to work. You will find yourself deliberately searching out and checking each item as the lists guide you around the unfamiliar airframe and cockpit.

These are items that are essential as you compose your Starting Engine checklist:

Radio and electrical accessories. Modern radio equipment can suffer damage if it's hit with a voltage surge. An alternator often develops this power surge during that moment when the engine first fires. For the protection of all radio equipment, see that all avionics are in an "off" position prior to starting.

Other electrical accessories should be subject to this same inspection. Aircraft batteries, for example, are built lightly to save weight. This means that they have limited cranking capacity. If many accessories are "on" during engine start, the battery will drain. While on the subject of batteries, a word here about hand proping an engine with a dead battery: It's dangerous business unless you have had instruction in the procedure. Furthermore, a battery that is too weak to crank the engine usually does not have the energy it takes to accept an in-flight recharge from the alternator. You will probably arrive at your destination with a still-dead battery.

Fuel selector. The busiest place in the sky is the cockpit of a Skyhawk with four pilots aboard when a tank runs dry. But imagine how all that enthusiasm and industry can quickly turn to frustration if the selector handle won't budge or comes off in the first hand to reach it! In preflighting, when you reach down to move the fuel selector to the tank required for takeoff, first rotate it through all positions.

Cowl flaps. Normally we pilots set the cowl flaps to "open" when we taxi so that there is an increase in the cooling airflow around the engine. In your engine checkup, cycle the cowl flaps to make sure the control will open and close during flight.

Carburetor heat. Make sure your carburetor heat is in proper position for taxiing. As a usual thing, taxi with the carburetor heat cold. In the "open" or hot position, unfiltered air is brought directly through the carburetor to the engine. If that air is laden with dust and grit you add unnecessary wear to the engine. Some engines, however, are prone to develop carburetor ice when taxied at low RPMs in humid air. If your engine does ice easily, taxi with the heater hot when there is visible moisture in the air.

Fuel mixture. Another preflight task calls for the proper fuel mixture position for engine start. A cold engine usually calls for a rich-mixture start. Hot engines, however, may crank more easily with a lean mixture.

Throttle and propeller. As part of the starting sequence, set the throttle and propeller as recommended by the handbook for starting. Avoid starting the engine with *excess* throttle. High RPMs during that moment after starting add unwarranted wear to an engine that has not quite yet coated its parts with oil. Also, be conscious of what lies in your prop wash. A blast of dust and grit aimed toward the maintenance hangar, for instance, is bound to stir an acquaintance with the mechanic—especially if he has an engine opened on the bench.

Engine primer. A properly primed engine cranks more quickly and saves wear on the electrical system and starter. In preparing for flight, use the manual-approved method of priming the engine; either the primer pump or the auxiliary fuel pump. Don't prime by "pumping" the throttle. This procedure is apt to flood the carburetor. A backfire on starting can produce a fire. If you should get a fire in a flooded carburetor, take steps to draw the excess fuel into the induction system: Close the mixture control and turn off the fuel pump; open the throttle and keep cranking. If the fire continues, of course, abandon the aircraft.

Master switch. Throwing the master switch in an airplane performs the same function as the first "notch" in your automobile's ignition switch—it activates the electrical system.

In many light aircraft, the master switch is a split-type switch with the halves labeled "battery" and "alternator." Normally both halves are used simultaneously. The alternator side of the switch may be turned off to remove the alternator from the electrical circuit while the circuit continues to function. It is standard procedure when starting the engine with external power in order to protect the alternator from possible surge damage. An erratic ammeter reading in flight is another reason to shut down the alternator. The electrical circuit will still function, but the entire electrical load is placed on the battery. That is why all nonessential electrical equipment should be turned off for the remainder of the flight.

Propeller area. Before starting the engine make sure the propeller area is clear. After you have yelled "Clear!" in a drill-sergeant octave, look left and right ahead, to the sides, and toward the rear. Look for those nonpilots who might think you are only announcing the fair sky. The time taken in search of a potential nasty accident prevents the shortest lapsed time in all of aviation—those microseconds that many pilots place between yelling the warning and spinning the blades.

For a night-time start, your preflight requires another step. If you are cranking the engine at night, be sure to flip on the navigation lights as you clear the propeller area. The small bulbs place only a small drain on the battery, and without their warning it is hard for a pedestrian on a dark ramp to tell which airplane is starting.

Starting the engine. If the engine is difficult to start, avoid prolonged cranking; the starter armature will overheat. If prolonged cranking becomes necessary, allow the starter to cool for ten seconds following each ten-second interval of cranking.

Oil pressure. As your final preflight check of the starting sequence, look at the oil pressure gauge. The needle on the gauge should move into the operating arc within thirty seconds of starting. If it does not, either the gauge is defective or the oil pressure is insufficient to lubricate the engine. Either problem could cause engine damage.

Before Takeoff

We never seem to outgrow our need for a preflight written checklist on every flight. An uncaring and impartial plane reminded me of this simple truth a few years ago. I thought at the time that I was through doing foolish things around an airplane, but the next six minutes proved me wrong on that score.

Like most charters, the flight was a hurry-up job. The FSS briefer gave me the weather while I scribbled out the flight plan and stuffed down an airport breakfast from the red, stale-cracker machine.

A cold, light drizzle drifted across the ramp when I reached the plane, one I had literally flown hundreds of times. The passenger hurried aboard. I shoved his oversized suitcase in after him, climbed into the cockpit, and fired up. I ran the pretakeoff checks from memory, out-taxied a Lear to the runup area, got clearance on the roll, swung onto the runway, and mashed in full throttle.

"Wow!" I thought as the plane accelerated to rotate speed. "That's the way to get things going . . . no foolin' around." Then the passenger leaned over and pointed. "What's that red thing stickin' outta your steering wheel?"

Awk! . . . Gasp! . . . (blush) . . . Aarrgh! The gust lock was still in the yoke and my checklist was still in my locker!

Well, the runway was long, the brakes were good—and the airplane made its point; we just never outgrow our need for a written checklist.

During World War II, pilots often shifted from one type of aircraft to another. They required a pretakeoff checklist that would work in any plane. The service devised a simple eight-letter reminder list: CIGFPTRS. Let's look at the eight factors represented by the letters.

Controls. By all means check the controls for full travel and

Controls
Instruments
Gas
Flaps
Prop
Trim
Run-up
Seatbelts

CIGFPTRS—an easy before-takeoff checklist that works in any plane.

feel for any binding. A word here about those "throw-over" yokes: If you anticipate the right-seat pilot possibly using the controls, rehearse a "throw-over" while you are still on the ground. By staging this rehearsal you will make sure the system is in working order and that each pilot knows the moves.

Instruments. Each and every gauge on the panel has its own proper indication as you sit in the runup area. Check the instruments in a logical sequence so that none is overlooked. The system I use reads the gauges from left to right, starting with the top row and checking all the way across the panel before dropping to the next row.

Here are some items to consider as you look at the instruments:

magnetic compass   Check the reading against a known direction (parallel or square with the runway, for example). See that the face is filled with fluid and that the compass card rests level. A whiff of alcohol or mineral spirits is a sure sign that the instrument is leaking and needs service.

airspeed indicator   The airspeed indicator needle should register zero with the plane at rest. Take a moment, as well to determine which is the nautical scale and which the statute.

attitude indicator (artificial horizon)   Adjust the nose dot of the attitude indicator so that it reads level as you sit in the runup area. (If you are flying a tail-wheel aircraft, however, the parked indication should read nose high.) Check the suction gauge at this time for vacuum. The careful pilot looks for any "banked" attitude on the instrument as he taxis around corners on his way to the runup area. Any "bank" in excess of five degrees means that the instrument is excessively worn and unreliable.

clock   Make sure your plane's clock is set and running. It is annoying to discover well into the flight, with the gas gauges dwindling, that the clock and your wristwatch are forty-five minutes out of agreement.

altimeter   Adjust the altimeter to the field elevation or turn the Kollsman to the current altimeter setting. With the Kollsman properly set, do not be too surprised to see a moderate disagreement between the altimeter reading and the published field elevation. The published elevation represents the highest point on a usable runway. Unless you happen to be sitting on that point, there is bound to be a discrepancy of several feet on even the most level airport. Remember, altimeters are calibrated by the manufacturers with the assumption that the average instrument panel sits fifteen feet above the runway. So don't be worried by

Checking the Plane for Flight

a small discrepancy; just set the Kollsman to the current altimeter setting and accept a reasonable "error." If you do find that the discrepancy is significant, say it exceeds seventy-five feet, then consider the altimeter unserviceable.

tachometer Even at idle power the tachometer needle should hold steady and be free from erratic movement.

turn needle and ball While you are parked on the runup apron, the turn needle should indicate a zero rate of turn. During taxi, take note that the needle indicates the proper direction of taxiing turns and that the ball moves freely *away* from the direction of the turns.

heading indicator (directional gyro) As a preflight duty align the heading indicator with the magnetic compass. Remember though, that the instrument required about two minutes "warm-up" after starting, for the gyroscope to attain stability and accuracy.

vertical speed indicator Even before flight the vertical speed indicator has a proper reading. Note the instrument reading as you sit at idle, for that value represents the *zero* in flight. If, for example, the needle rests on 100 FPM "down," then that registers level flight, and a descent of 300 FPM will show up on the dial as 400.

fuel gauges As you know, fuel gauges cannot be depended upon to register the exact amount of gas in the tanks. Their indication before flight, however, should approximate the known quantity in the tank.

engine temperature As part of your "before-takeoff check," wait until the engine temperature reaches the green arc before you attempt the takeoff. Otherwise the engine may run rough and not deliver full power on the takeoff run. On cold days this may mean a few minutes of warm-up. In this case, avoid warming the engine at very low idle RPMs. Avoid spark plug fouling which robs the engine of full takeoff power, with a warm-up of 1,000 to 1,200 RPM.

ammeter Some ammeters display the electrical load being placed against the alternator, while others are designed to show whether the battery is in a stage of *charge* or *discharge*. Review at preflight the aircraft owner's manual to determine the normal indication for the plane you fly.

Many light airplanes are also equipped with an alternator-malfunction warning light. If a malfunction occurs, the light illuminates. If the condition occurs in flight, reset the voltage relay by momentarily turning the master switch to *off*, then back to *on*. Should the malfunction persist, turn the alternator position of the master switch off and conserve electrical consumption since only battery power is provided.

Gasoline. In most light airplanes, there are four fuel-related items to check prior to takeoff:

fuel pump Momentarily turn off the electrical boost pump, to verify that the engine-driven pump is working. Then return the boost switch to its takeoff position. As a general rule I use the electric pump anytime that my plane is within a thousand feet of the ground. I feel that this practice does not add unwarranted wear to the boost pump, and it protects against a dangerous power loss, should the engine-driven pump fail at low altitude.

mixture At normally anticipated field elevations, a full-rich mixture helps cool the engine during the full-power takeoff. But if the density altitude exceeds 5,000 feet at the departure airport, some leaning is needed for takeoff to prevent a rough-running engine. At higher altitude airports, run the engine to full throttle for a few moments while you lean the mixture just as you would in cruise flight.

selector handle It is necessary before flight to set the selector handle at the proper position. In some airplanes all tanks will not feed properly in a lift-off attitude or steep turn. So if the aircraft handbook calls for a certain tank for takeoff, use it.

quantity Make sure that the selected tank holds at least the minimum fuel required for takeoff. As I have previously pointed out, do not count on an accurate reading from the fuel gauge, and as a yardstick do not attempt a takeoff with less than a quarter of a tank showing on the instrument. There are fewer tight spots in flying than a fuel-starved plane right after lift-off with no landable patch of survival within reach. Injury to pilot and passengers is almost certain.

Flaps. Fully extend and retract the flaps to test for operation, and look for any binding or erratic movement. Then set the flaps for takeoff, as recommended by the manufacturer.

Propeller. If you are flying a plane with a controllable propeller, test the pitch control before takeoff. Set the power at 1,700–2,000 RPM (adequate revolutions to operate the pitch control) and move the blades to *low RPM*. Proper operation is indicated by a 100–400 RPM drop. On cold days you may need to exercise the control three or four times to obtain the drop in RPMs. Once the test is satisfactorily completed, return the propeller control to *high RPM* so the engine can accept full throttle on takeoff.

Trim. An out-of-trim airplane holds the potential for a deadly takeoff. Imagine a few common situations that can easily distract a pilot on lift-off: A door pops open, his seat slides to the rear stop, or the passenger's feet entangle the rudder pedals. Then picture the confused pilot momentarily ignoring his primary job—flying the airplane—to correct the situation. If the plane is trimmed nose high, a stall is a very real possibility; nose low trim and a dive into the ground is a likely prospect. Rudder trim is equally important. Adverse rudder trim, of course, would probably lead to a turn and then a dive. A properly trimmed airplane gives a distracted pilot a few moments to gather his wits. Before you set the trim for takeoff, check the control for freedom of movement. Exercise both the elevator trim and rudder trim to their maximun deflections.

Runup. When checking the magnetos individually, test the mag position *farthest* from the "both" position first. This helps to prevent a pilot from inadvertently launching on one magneto only, after he has finished the checks. It is difficult to visually determine if the key has been returned to "both" or the first magneto only. But if that first magneto position is checked last, this mistake cannot easily happen.

When you apply carburetor heat during the runup, you are checking for the existence of ice as well as for the proper functioning of the carburetor heater. Induction icing can easily form during prolonged taxiing at minimal power. But a carburetor heat test at runup RPMs quickly discloses any ice by an excessive power drop and rough engine. Needless to say, you want to clear the carburetor before you commence your takeoff.

Seat belts and doors. Personally determine that each passenger has fastened his seat belt and shoulder harness before you move from the runup area. Make sure all doors are closed and locked. And if a door should pop open during takeoff, know that there is no cause for alarm; it will not adversely affect the plane, and the slipstream will keep it tight against the fuselage. If the runway is long and the door opens on the ground run, simply brake to a stop, shut the door, and taxi back for takeoff. If the door opens after you're airborne, you have two choices. Either circle back and land, or close it after you reach your altitude. But in no instance should a pilot wrestle with a door at a low altitude.

The checklists that follow the chapter review are designed to accommodate most modern light aircraft. The plane you fly, however, may demand special additions to the list. Refer to your aircraft operations manual for these special preflight requirements and make your additions in the space provided.

In Review

- Conducting a proper preflight inspection depends on the pilot's willingness to expend the necessary time and effort, the use of an adequate written checklist, and the pilot's attitude toward preflight activities.
- Get to the airport well ahead of your ETD so that you do not rush through the preflight inspection.
- If you find that you are skipping over some checklist items, or checking some out of sequence, be warned that your attention has wandered. Start your inspection over.
- Guard against the fallacy of inspecting an airplane with the assumption that it is probably OK.
- A pilot who was not taught how to properly preflight an airplane *must* take the time to teach himself; he cannot continue to let chance see him safely through his flying career.
- A competent preflight inspection does not depend on the pilot's familiarity with the plane before him; it depends on his familiarity with his own checklist.
- We never outgrow the need for a written checklist on every flight.

Preflight Aids

I. Subject Area: Aircraft Preflight Inspection

   A. Before Inspection Begins
   1. Switches off and remove key
   2. Remove gust lock
   3. Set parking brake
   4. Gear handle down and locked
   5. Remove tie-downs or chocks
      other _____

   B. Wings
   1. Ailerons—check
   2. Fuel sumps—drain
   3. Pitot tube—check
   4. Stall indicator—activate
   5. Fuel tanks—check quantity; caps secure
   6. Wing tip and nav light—check
      other _____

   C. Landing Gear
   1. Tire wear and inflation—check
   2. Brakes—check for fluid leak
   3. Wheel-well door—check
   4. Retract mechanism—check
   5. Wheel struts—check
      other _____

   D. Nose Section
   1. Windscreen—clean
   2. Cowl flaps—check
   3. Propeller and spinner—check
   4. Induction air intake—unobstructed
   5. Air inlets—check for foreign matter
   6. Oil level—check; cap secure
   7. Cowl interior—check for oil/fuel leaks

Checking the Plane for Flight 159

    8. Exhaust system—check
    9. Wiring/belts/hoses—check condition
   10. Fuel strainer—drain and secure
   11. Engine inspection covers—secure
       other _____

E. Empennage
    1. Control surfaces—check
    2. Drain ports—open
       other _____

F. Fuselage
    1. Static pressure port—check
    2. Tow bar—stowed
    3. Emergency locator transmitter—armed
    4. Fire extinguisher—charged; secure
    5. Baggage door—secure
    6. Exterior and interior lights—check
    7. All surfaces—free of frost; check for abnormalities
    8. Aircraft documents and aircraft inspection date—check
       other _____

II. Subject Area: Starting Engine

  A. Radios and Electrical Accessories—off
  B. Fuel Selector—test for freedom of movement; proper tank
  C. Cowl Flaps—open
  D. Carburetor Heat—open
  E. Mixture—full rich
  F. Propeller—high RPM
  G. Throttle—start position
  H. Prime Engine
  I. Master Switch—on
  J. Propeller Area—"Clear"
  K. Magnetos—on
  L. Start Engine

M. Oil/fuel Pressure—check
N. Radios and Electrical Accessories—as required
O. Warm-up—800 to 1,200 RPM
P. Parking Brake—release
   other _____
Q. Notes for Flooded and Hot Starts
   1.
   2.
   3.
   4.

III. Subject Area: Before Takeoff

   A. Parking Brake—set
   B. Controls—full movement
   C. Instruments—check and set
   D. Gasoline
      1. Fuel pump—test, then on
      2. Quantity—above takeoff minimum
      3. Selector—verify proper tank
      4. Mixture—takeoff setting consistent with field elevation
   E. Flaps—test cycle and set for takeoff
   F. Propeller—exercise
   G. Trim—test for cycle and set for takeoff
   H. Runup—check magnetos/carburetor heat
   I. Seat Belts and Harness—check passengers
   J. Doors and Windows—secure
   K. Parking Brake—release
      other _____

CHAPTER 7: **Complying with FARs**

THERE ARE ELEVEN REGULATIONS that a pilot should review before each cross-country flight. Make an initial reading of these FARs in their entirety for a full understanding of their content. Then use this synopsis as a preflight guide to comply with the Federal Aviation Regulations.

FAR 91.22 FUEL REQUIREMENTS FOR FLIGHT UNDER VFR

By regulation a VFR pilot cannot commence a cross-country flight without a prescribed fuel reserve. For daylight flying he must depart with enough fuel to reach his destination plus thirty minutes endurance at cruise power. A forty-five-minute reserve is required for night flight. Are these FAR minimums adequate for safety? That depends largely on four factors:

1. The speed of the airplane
2. The availability of alternate airports
3. The weather
4. The pilot's familiarity with the terrain

Obviously, a 170-knot Bonanza is able to range farther in the reserve time than a 90-knot Cessna 152. And extra range means that more alternate airports are available to the pilot.

The farther apart the airports along your route, the more gas it may take to reach one of them as an alternate. Don't count too heavily on finding the unpaved airports if you don't know the territory. And remember to disregard at night all those that do not have a rotating beacon.

If en route weather is anticipated, count on detours that eat up a lot of your reserve before you gain your destination. If any weather exists at your destination, you may need to fly a zigzag course to an alternate airport.

If you are unfamiliar with the terrain around your destination, cover that possibility of getting just a little bit lost with an extra fuel reserve. After all, a half hour's fuel only registers an eighth of a tank in most light planes—and that's a terrible sight to see on the instrument panel when you don't know exactly which way to turn.

I have always limited my time aloft to allow a landing with at least an hour's gas on the clock and a quarter tank on the fuel gauge. I have never been sorry for the practice and I have not begrudged the extra fuel stops it has cost me.

FAR 91.23 FUEL REQUIREMENTS FOR FLIGHT INTO IFR CONDITIONS

An IFR pilot's fuel reserve must let him fly to his destination, then carry him onward to his alternate airport, to arrive there with forty-five minutes still in the tanks.

This fuel requirement may be shortened, however, if three conditions are met by the destination airport:

1. It has a published instrument approach.
2. The ceiling is forecast to remain at least 2,000 feet from one hour before your ETA to one hour after your ETA.
3. Visibility is forecast to remain at least three miles from one hour before ETA to one hour after ETA.

Then, only fuel for the destination plus forty-five minutes is required.

FAR 91.24 ATC TRANSPONDER AND ALTITUDE-REPORTING EQUIPMENT

In brief, this regulation requires a pilot to use a transponder with altitude-reporting capability anytime he flies above 12,500 feet or within a Terminal Control Area. ATC usually admits a nonaltitude reporting transponder just for the asking and, given advance notice, often allows a plane with no transponder at all.

FAR 91.25 VOR EQUIPMENT CHECK

Before filing an IFR flight plan, make sure the VOR equipment has received an accuracy test within the past thirty days. The test must be conducted by the pilot and entered into the aircraft logbook. The log entry must show the date of test, place, and degree of error.

There are four permissible methods for determining the accuracy of the airplane's VOR receiver:

VOT test signal:	±4 degrees allowed
VOR checkpoint:	±4 degrees allowed
Airborne check:	±6 degrees allowed
Dual VOR:	±4 degrees allowed

FAR 91.27 AND 91.31 CERTIFICATIONS REQUIRED

These two regulations spell out the four documents that must be aboard any aircraft. Any pilot renting or borrowing an airplane must see that these are in the cockpit. To recall the required papers easily, just remember the word "WRAP":

Weight and balance data
Registration of ownership
Airworthiness certificate
Placards of operating limitations

FAR 91.29 AIRCRAFT AIRWORTHINESS

The regulation simply says: "The pilot in command of a civil aircraft is responsible for determining whether that aircraft is in condition for safe flight."

FAR 91.32 SUPPLEMENTAL OXYGEN

The pilot must use supplemental oxygen anytime he flies above 12,500 feet for more than thirty minutes. (If above 14,000 feet, however, he must go on oxygen immediately.) Additionally, each passenger must have oxygen available to him at altitudes of over 15,000 feet.

FAR 91.169 INSPECTIONS

This FAA rule states that privately owned and operated aircraft must have an annual inspection by an FAA improved inspector. Commercially flown or rented aircraft are required to be inspected every one hundred hours as well as have an annual inspection. In each case, compliance with the rule is the responsibility of the pilot.

FAR 91.170 ALTIMETER INSPECTION

For IFR operations, an instrument test within the preceding twenty-four months is required.

FAR 91.177 TRANSPONDER INSPECTION

A test within the preceding twenty-four months is required.

CHAPTER 8: **Qualifying the Plane for IFR Flight**

THE PILOT preparing for an instrument cross-country has several more aircraft preflight items to consider than has the VFR pilot. Those items concern evaluating the aircraft's equipment in light of the specific IFR needs. Several details of aircraft equipment demand special preflight checks. An evaluation of the plane's radio equipment, for example, is of prime importance when preflighting for an instrument cross-country. This importance stems from the two basic functions that a radio serves to instrument flights: (1) once into IFR conditions, the pilot's clear communication with the controller is his primary means of avoiding other aircraft, and (2) with the instrument conditions hiding the terrain's landmarks, the VOR becomes the pilot's primary mode of navigation.

Since radio communications and navigation are so indispensable to instrument flying, a preflight evaluation of the plane's radio equipment is a major concern to the departing IFR pilot. Call the ground controller and ask for a radio check during preflight to insure quality transmission. The ground controller will then advise you as to the strength and quality of the radio communications. He will normally use the term "loud and clear" for a transmission that has adequate strength and quality. If the signal is strong but distorted, the controller might advise "loud but garbled." If the signal is weak but the words are distinct, he will advise "weak but clear." "Weak and garbled," of course, means the signal volume is low and the words are indistinct.

You should hesitate to go IFR with a radio that checks out as either "weak" or "garbled." Your IFR procedures are just too dependent on good communications. A trip to the radio shop may be needed, but first try a few simple cockpit remedies. A mike held more than a couple of inches from your mouth, for example, may be the cause for a weak broadcast. The combination of an open window and engine noise can account for a garbled transmission. A few other simple reasons for an "inoperative" radio include:

Talking into the back of the mike
Keying the wrong frequency
A partially loosened mike plug
A fully squelched receiver
The volume turned down
The wrong toggle switch on the audio
    selector panel

Look for these symptoms before heading for the radio shop.

In addition to preflighting the strength and quality of his radio transmitter, the departing IFR pilot should evaluate the accuracy of his VOR receiver. He should conduct an accuracy test before each flight even though the FARs only require a check within the thirty-day period prior to the flight. A check prior to each flight is particularly important when a pilot is planning to put a rented or borrowed aircraft into the IFR system. In these cases the departing pilot has no idea where the quality of the plane's radio equipment stands. A preflight test of the VOR receiver is the only assurance a pilot has that his primary IFR navigation system is flight-ready.

The pilot has a choice of three preflight methods for testing his VOR. The first and most accurate method employs the VOR test signal (VOT) that is available at many large airports. (The availability of this test signal and its frequency is listed by airport in the back of the *Airport/Facility Directory*.) The VOT signal is nondirectional in nature and can be used from any position on the airport. Its accuracy is guaranteed *only* when used as a ground test; it may be subject to interference when received by an airborne plane.

To conduct the VOT check, first tune to the test-signal frequency given in the *Airport/Facility Directory*. Then turn your VOR's omni bearing selector (OBS) until the VOR needle is centered. If the "to-from" window reads "to," the OBS must read within four degrees either side of "180" for an accurate VOR. If the "to-from" indication is "from," the OBS must read within four degrees of "O."

Many airports without a VOT facility instead offer a designated VOR checkpoint, also listed in the *Airport/Facility Directory*. This checkpoint is normally marked on a ramp area and lies on a known radial from the VOR station. Directions for finding the point are given in the *Airport/Facility Directory*, along with the certified radial that crosses the checkpoint.

To conduct a preflight VOR test using a designated checkpoint, taxi to the point and tune your receiver to the VOR frequency. Then set the OBS to the certified radial. The "to-from" window should read "from" with no more than four degrees correction required to center the needle with the OBS.

If your departure airport offers neither a VOT facility nor a designated checkpoint, preflight your VOR's accuracy by means of the "dual system check." To use this system, merely tune both of your VOR receivers to a nearby VOR station. Then center both needles with "from" readings. Allow only a four-degree variation between the two indicated bearings. (This system assumes that the IFR pilot has two VOR receivers aboard, of course—and for safety's sake he really needs that second radio.)

In addition to evaluating the airplane's radio installation, the prudent IFR pilot carries aboard his own "backup" radio components: a spare mike and a set of earphones. Microphones and cabin speakers seem to pick the worst possible moment to quit working. And at that moment the pilot is either glad that he brought spare equipment along or sad that he did not. The pilot who brings the spare mike and earphones along, however, must also remember that spare equipment that doesn't work is of little value. He should, therefore, check his spares, as he preflights his plane's installed radio equipment.

Aircraft radio installations are not standardized. Additionally, these installations are often complex, particularly in older planes that have had their radio components changed and upgraded. A pilot preparing to fly a rented or borrowed aircraft must devote additional preflight time to a review of the plane's avionics layout. He cannot safely enter the IFR environment until he is instinctively familiar with the locations and manipulations of the audio selector panel, transmitter selector switches, Nav/Com function selectors, and the existence of any "series" radio hookups. ("Series" here means that a particular radio must be on for other radio equipment to operate.) A tired and distracted pilot unfamiliar with the aircraft's radio installation can easily make the wrong move at the wrong time. (I am reminded of a pilot who was taking some recurrent training in an unfamiliar plane. We were still in the clouds following the localizer down final approach and descending through four hundred feet when he reached over to quieten the annoying *beep-beeeep* of the middle marker. Instead, he hit the wrong toggle switch and doused *all* communication and navigation. The first few moments of the ensuing missed approach were, ah . . . interesting. By the time the elusive "on" switch was found, and the radios started talking and navigating again, the controller was already well into his loud list of questions.)

In addition to complex radio installations, the older plane often presents another possible source of trouble for the instrument pilot—a generator of inadequate capacity. Many of the older planes are equipped with a thirty-five-amp generator. This worked fine with the limited electrical accessories and minimal radio equipment that was in vogue when the plane was manufactured. Years of upgrading the avionics and electrical equipment, however, often increases the electrical drain beyond the capability of the old generators. If the plane you plan to fly IFR belongs to a previous decade, check the total electrical drain described on the aircraft's equipment list. (You may delete the radio transmitter drain, as it is used only intermittently.) To evaluate the plane for instrument operations, compare the electrical system's total drain with the generator's output.

As another preflight evaluation of the aircraft equipment, verify that an alternate static source is provided. Instrument conditions may produce the factors that lead to a plugged outside static vent. This condition renders the airspeed indicator, altimeter, and vertical speed indicator unreliable. The loss of these primary instruments is intolerable in IFR conditions, unless the pilot can get them working quickly again by switching to the airplane's alternate static source switch. (Unfortunately there is no reliable preflight test to assure that the alternate static source itself is in proper working order.) The instrument pilot, however, may want to extend his preflight inspection to those first few minutes after he is level on his cruising altitude. Then, while still VFR, he can check the condition of his alternate static vent by momentarily turning its switch on. If the system is working, he will note a small increase in both the airspeed indication and the indicated altitude. Typically, the airspeed indication increases about five knots; the altimeter about fifty feet.

As a final evaluation of the aircraft's equipment, the departing instrument pilot should consider the need he may have for an autopilot with at least "wing-leveling" capability. If the pilot does not fly IFR on a regular basis he may tire quickly in IFR conditions. Even the simplest autopilot system will give him the helping hand he may need if the flight is long, the instrument conditions extensive, and the traffic heavy. The autopilot will at least hold a steady course for him while he tends to his communications and navigation. While these autopilots vary in operation, most outline a preflight test for the systems reliability in the operator's handbook. The IFR pilot should incorporate this handbook test in his preflight evaluation of the aircraft's equipment.

A copilot is an alternative to an autopilot. Just make sure that he is IFR competent. A competent copilot can offer invaluable help with communications and navigation. A copilot who is unfamiliar with instrument procedure, however, is a liability in the cockpit. You must monitor his actions and cover his mistakes in addition to your normal cockpit work load.

In Review

- The IFR pilot must extend his preflight inspection of the airplane to include the specific needs for instrument flying.
- An evaluation of the airplane's radio equipment is of prime importance.
- The radio serves two basic IFR functions.

    1. Aircraft separation through pilot/controller communications
    2. Navigation through VOR reception

- Ask the ground controller for a preflight evaluation of the radio transmitter's strength and quality.

    "Loud and clear" means the signal strength and quality are both satisfactory.
    "Loud but garbled" means the signal volume is strong but the words are distorted.
    "Weak but clear" means the signal volume is low but the words are distinct.
    "Weak and garbled" means that the signal strength and quality are both unsatisfactory.

- A mike held too far from the mouth may be the cause for a weak signal, an open window the reason for a garbled transmission.
- Other common errors that make a radio seem inoperative include:

    Talking into the back of the mike
    Keying the wrong frequency
    A loosened mike plug
    A fully squelched receiver
    The volume turned down
    The wrong toggle switched on the audio selector panel

- A pilot should not go IFR with a radio that checks out as either "weak" or "garbled."
- The preflight inspection of the radio equipment should include a VOR accuracy check by one of the following:

    A VOT signal
    A designated VOR checkpoint
    A dual system check

- A preflight check of the radio equipment should include a verification that the spare mike and earphones are in working order.
- The preflight inspection of a borrowed or rented aircraft must include a review of the plane's avionics layout.
- If the plane you plan to fly IFR is an older model, check the total electrical drain (excluding the transmitter) described in the aircraft's equipment list. To evaluate the plane for IFR operations, compare this drain with the generator's output.
- Verify that the plane you intend to fly IFR provides an alternate static source.
- Even the simplest autopilot will give you a helping hand if the flight is long, the IFR conditions extensive, or the traffic heavy.

Preflight Aids

IFR Preflight Equipment Check

1. Confirm strength and quality of radio communication.
2. Evaluate accuracy of VOR receiver.
3. Check condition of spare mike and earphones.
4. Review avionics layout.
5. Compare electrical drain to generator output.
6. Confirm that the plane has an alternate static source.
7. Check the autopilot for proper operation.

VOR Accuracy Tests

VOT TEST

1. Refer to *Airport/Facility Directory* for frequency.
2. Turn OBS to center the needle.
3. With "to," the OBS must read $\pm 4°$ of "180."
4. With "from," the OBS must read $\pm 4°$ of "0."

DESIGNATED CHECKPOINT TEST

1. Refer to *Airport/Facility Directory* for location of checkpoint and certified radial.
2. Position the aircraft on the checkpoint.
3. Set OBS to the certified radial.
4. With "from," the needle should center within $4°$ of OBS retuning.

DUAL SYSTEM TEST

1. Tune both VORs to nearby station.
2. Center both needles with a "from" reading.
3. Allow only a $4°$ variation between indicated bearings.

THE ENVIRONMENT

The airplane has unveiled for us the true face of the earth.
SAINT EXUPÉRY

CHAPTER 9: Relating the Weather Briefing to the Flight

FREQUENTLY AT FLIGHT SERVICE STATIONS you see a pilot listening to the weather briefer across the counter. After the briefing is finished, the pilot turns and walks away, and the expression on his face says: "Now, what did he say?"

The reason a pilot sometimes misses the spoken word at an FSS briefing is seldom due to a complicated weather pattern. It usually happens because the pilot goes into the briefing cold; he has not prepared himself to hear the expected. As an example of this, picture a person walking toward you on the street. He looks questioningly at the watch on his wrist, turns to you and starts to speak. Now, if he asks, "Do you have the correct time?" you hear every word—you were prepared to hear the expected. But if that person asks, "Are the slopes firm in Sun Valley?" you don't hear him, because you went into the conversation cold.

And so it is with understanding the weather briefer. The first step, then, is to prepare your logic for what it expects to hear. And the best way to accomplish this is by maintaining a constant awareness of aviation weather. Make yourself a pilot even on those days between flights. Watch the nightly TV weather through the eyes of a flier. If FSS has automatic telephone weather service in your area, make daily calls for the forecast. Stay in tune with the changing weather patterns on a day-to-day basis. (You will know that you are completely in tune when your first action of the day is to pull aside the curtains to see how your sky has fared the night.)

Then when you talk to the briefer you already know the big weather picture; you are prepared to hear the expected. You need now only to fill in the details.

Whenever possible I like to visit FSS in person to get my briefing. And before the briefing begins, I ask to see three charts that give a quick overall weather picture: (1) the surface analysis, (2) the weather depiction, and (3) the radar summary chart.

The surface analysis (also called surface weather map) gives me a ready location of any front near my proposed flight and its direction of movement. Fronts do not always contain significant weather, but if there is any weather knocking about, it is usually along that frontal line.

The weather depiction chart is prepared from surface observations and outlines the areas of IFR (visibility less than three miles; ceilings below 1,000 feet) and marginal VFR (visibility five miles or less; ceilings 3,000 feet or less). If my course passes through these outlined areas, I question the briefer on the specifics, ask for recent pilot reports, and review current station reports in the outlined areas.

The radar summary chart displays areas of actual precip. It shows the type of weather (drizzle, rain, thunderstorms), the intensity, and whether that intensity is increasing or decreasing. The percentage of terrain coverage is given, along with the area's speed and direction of movement. If it appears that the area of radar weather echoes will cross my path, I definitely obtain current pilot reports. And since the weather pattern of an area of precip can change rapidly, I call back for any further pilot reports just before takeoff.

Notice that my prebriefing review favors *reports* of actual weather, rather than *forecasts* of expected weather. If I have the

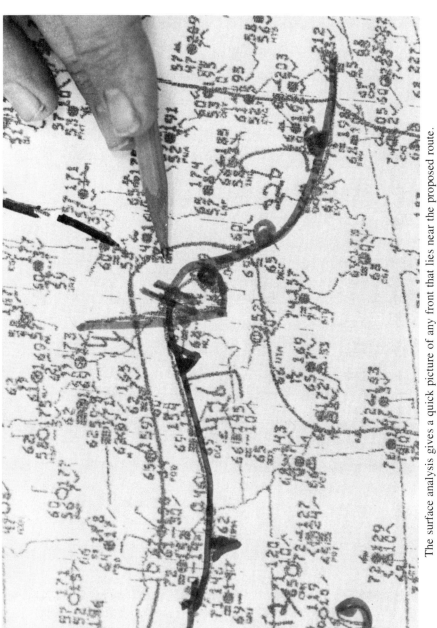

The surface analysis gives a quick picture of any front that lies near the proposed route.

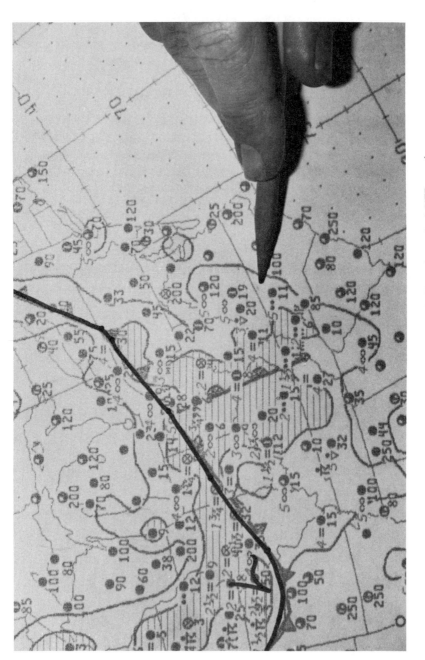

The weather depiction chart outlines areas of IFR and marginal VFR weather.

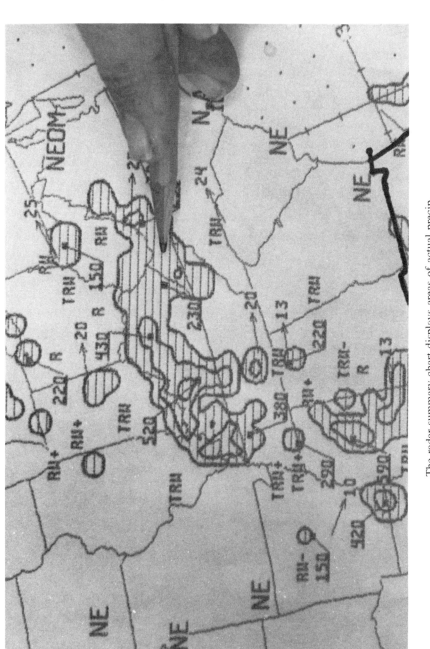

The radar summary chart displays areas of actual precip.

advantage of getting a weather-trained briefer to review the data with me, I prefer to hear his opinion on the forecasts. Most briefers, by the way, are well schooled by the time they reach the flight service station. But they do come in all grades, just like the rest of us. So if you should land a briefer you feel is below par, don't hesitate to seek out another briefer for a second opinion.

When I ask for a weather briefing, I give the briefer three statements that avoid the necessity of his playing Twenty Questions before he starts reviewing the weather with me:

1. Whether I am flying VFR or IFR
2. My destination, route, alternate, and desired altitude
3. My proposed departure time, time en route, and any intermediate stops

Then as the briefer begins I listen attentively. I ask questions about items I do not understand, and reduce his spoken words to a written weather log—an invaluable cockpit reference. With it you can evaluate his reports and forecasts against what you see through the windscreen, and determine the trend—improving or deteriorating.

As the briefer and I bend over his materials on the FSS counter, I mentally relate the weather factors he predicts to the needs of the flight at hand. My first priority, of course, is whether or not the weather meets my personal weather preflight minimums.

If it does and the flight is on, I relate the briefer's information to the effects it will have on my cruising altitude, the route I take, and the refueling stops I make.

Altitude

A number of factors influence a pilot's selection of cruising altitude. First, significant *winds aloft* can certainly take an alti-

FACTORS THAT INFLUENCE PREPLANNED ALTITUDE

1. WINDS ALOFT

2. FREEZING LEVEL

3. CLOUD BASES

4. TURBULENCE

5. AIRCRAFT CEILING

6. TRANSPONDER EQUIPPED

tude away from you. If you pick the wrong altitude and catch a stiff head wind, the difference between fast and slow is whether the cars below are passing you up. If your head wind is strong you can easily estimate the percentage it will reduce your plane's range. Take the head wind velocity as a percentage of your plane's cruising speed. Then reduce your plane's "no-wind" range by that percentage. If, for example, your 120-knot airplane is facing a 30-knot head wind component, then its range suffers 25 percent. If its no-wind range is 400 miles with reserve, better figure on traveling just 300, or pick another altitude and wind aloft.

The second factor in selecting a cruising altitude concerns *freezing level*. Mix visible moisture, near-freezing temperature, and an airplane, and you get structural icing. Take any of these factors out of the mixture and you won't have ice. You can fly in visible moisture, such as a cloud or rain, with the outside air temperature gauge reading above "0" and the iceman stays away. (Be mindful, though, of the fact that aerodynamic cooling can reduce the air over the airfoil another two or three degrees.) Similarly, if it's freezing outside but the air is clear, just turn the cabin heat up to "9" and forge ahead. But if the chunk of sky is both wet and freezing, then you must remove the plane.

When selecting your altitude, you must relate any freezing level to known areas of precip or cloud levels. Forecasts of freezing levels are fairly accurate, but the geographic area is somewhat harder to pin down, particularly if precipitation is involved. This is due to the changeable nature of precip. It moves and changes in intensity; one area of rain lets up, just to start in another.

The freezing-level forecast usually describes the type of ice expected—rime or clear, with clear ice the more hazardous of the two. In actual fact the pilot is apt to encounter mixed icing in forecast rime ice areas, and that brand forms rapidly.

The briefer speaks in terms of icing intensity—trace, light, moderate, and severe. These intensities refer to the rate of buildup that the pilot can expect on his airplane. A trace forecast tells us to expect a slow rate of accumulation that will not present a hazard within one hour of exposure, even without deicing or anti-icing equipment. A light forecast tells us accumulation may create a problem after an hour's exposure but may be controlled with de-icing or anti-icing equipment. A moderate forecast warns us that the icing is dangerous for even short encounters and requires deicing or anti-icing equipment. A severe forecast implies an immediate danger that even deicing and anti-icing equipment cannot defend against.

The problem with these intensity definitions is that the ice predicted very rarely follows. A pilot climbing toward trace is very apt to encounter moderate. A pilot flying in light may hit cold blasts of severe. Further, while the forecast of freezing level may be accurate, the intensity is difficult to define. This is not the fault of the forecaster; icing is just a highly fickle condition.

Many pilots who live in areas where they do not often encounter ice do not realize the perils associated with it. Foremost, ice adds weight. It also robs the plane's performance and capabilities in ways other than sheer added weight. Less than half an inch on the wing can reduce the lift 50 percent or more on some light planes, while it adds at the same time a similar percentage to drag. Thrust also suffers when ice interrupts the efficiency of the propeller blades, and a ring of rime in addition constricts the carburetor air intake. Even a coating of ice on an antenna can set up a vibration that blocks communication and navigation. Further, when Pitot tubes or static vents are glazed with ice, altimeter, airspeed indicator, and vertical speed indicator all become unreliable. What's more, any water that freezes in the crevices of control surfaces, wheel wells, and brakes delivers its own brand of excitement when it comes time to land—particularly with a frosted windscreen.

The pilot (usually IFR) who suspects that he may encounter ice should take a few defensive preflight steps. His weather briefing should determine which way he must fly to reach clear air along his route and whether he must climb or descend to warmer air. A current pilot report is invaluable, and the pilot must keep these reports updated once aloft. If he is one of the lucky light-plane pilots with deicing or anti-icing equipment aboard, a preflight review of the instructions for proper use is in order. Systems vary and improper operation can render them useless. I suggest checking the plane's manual and testing all available equipment. Any pilot who flies in ice should make the decision that he will break out of the ice conditions the instant structural ice begins to worry him. It can build fast. A pilot who delays until he can't hold altitude has just about run out of options.

I treat a known icing condition with the same degree of awe I treat a known thunderstorm. Whether advertised as weak or strong, I preflight-plan my best to stay out of it.

Cloud level is the third element that restricts the pilot's usable altitudes. The pilot must correlate cloud levels with terrain clearance, his own criteria for cloud clearance if VFR, and freezing levels if IFR. Clouds often usurp the allowable VFR cruising altitude you intended to fly, and you may be forced a flight beneath VOR and communication range. (The Low-Altitude Route chart of the instrument pilot helps the VFR flier. The airways have reception altitudes printed on them in the form of minimum en route altitudes.)

Turbulence is the fourth factor of the weather briefing that may narrow a pilot's choice of altitude. In preflight considerations, think in terms of passenger comfort and aircraft stress and then apply your own personal weather criteria. Be aware, however, that predictions of turbulence are similar to predictions of icing conditions, that is, the area of turbulence may move from its advertised position and change in intensity. If this is the case, pilot reports are in order and should be updated as you fly your route.

Naturally, factors other than weather affect your preflight selection of cruising altitude.

Aircraft capability is obviously a factor; most light planes start losing performance around 10,000 feet.

Transponder equipment is another factor. If the plane is not transponder equipped, flights above 12,500 feet might result in a case of your flying against the regulations.

Oxygen availability is still another factor. For flights above 10,000 feet (8,000 at night), the average pilot should have oxygen available. (Heavy smokers should reduce these figures by a fourth.) Hypoxia is a sneaky ailment that can come to a pilot without any real warning. We tend to think of hypoxia (the lack of sufficient oxygen) as a malady that causes the pilot to keel over at the controls. But it is usually not that dramatic. What it does is slightly impair the pilot's reaction time, judgment, and eyesight; just enough perhaps, to invite pilot error. When I fly above 9,000 feet without oxygen, I set a "hypoxia alarm" for myself. Knowing that vision is the first faculty to suffer, I look at the smallest thing I can read on the panel—for me, the Kollsman window. When those figures blur, I know it's time to descend a bit.

A final factor is *passenger discomfort*. Don't overlook it. Ailments of the ear or sinus, respiratory problems, or simply age can make higher altitudes very uncomfortable to some passengers and highly distracting to a pilot with a full-time job of flying the airplane.

Route

Once your preflight work has dealt with altitude considerations, the route is the next area to focus upon. Areas of precipitation that cover a percentage of land area greater than your personal weather minimum constitute reason to reroute your

FACTORS THAT INFLUENCE PREPLANNED ROUTE

1. EXTENT OF PRECIP.

2. POSITION OF FRONTS

3. ALTERNATE AIRPORTS

4. AREAS OF HEAVY TRAFFIC

flight. If an area of scattered thunderstorms is scheduled to cross your path, take a close look at the radar summary chart. If the echoes are not separated by at least forty miles, you will have difficulty circumnavigating them, even if good VFR exists between the storms. IFR flights are concerned with even isolated thunderstorms if they are "embedded" (hidden by surrounding clouds). Normally, airborne radar helps you avoid these storm cells and makes the flight possible. But if the cells are closer than the necessary forty-mile separation, then even radar cannot insure a safe journey. Remember, radar only allows a pilot to avoid an area of close-packed thunderstorms, not to penetrate it.

In addition to thunderstorms, areas of low visibility, turbulence, or low clouds that do not meet your criteria for acceptable weather minimums are other reasons for altering your preferred routing. Fronts are usually a factor in planning a route. Even if fronts contain little weather, frontal conditions can change rapidly. If you do plan a flight along a front, make certain that your route takes in sufficient alternate airports should they be quickly needed.

Apart from avoiding weather, the availability of alternate airports is a factor to consider when planning your route. If a night flight is contemplated, you may want to bend your route to keep you over the lighted areas of large highways and cities. Those towns that show up as yellow patches on the sectional chart usually display enough light to aid the nighttime VFR pilot. The shape of that yellow patch, by the way, represents the shape of the lighted area as it appears from the air. Any pilot who tries to avoid areas of heavy traffic in his car will probably do the same in his plane. Some sources of heavy air traffic en route are: large cities; military operations areas (MOA); the airspace surrounding a restricted area; coastlines; two or more controlled airports that lie within fifteen miles of one another; terminal control areas (TCA) and terminal radar service areas (TRSA).

Refueling or Rest Stops

Not only the route but also refueling or rest stops are affected by the weather briefing and require preflight attention. One factor that merits pilot consideration is *aircraft capability*. Winds aloft can either shorten or lengthen a plane's range and capability dramatically. For example, if a pilot has to weave his way through an area of rain he can count on more frequent refueling. Not only is his range shortened by the zigzags, but he also needs a greater fuel reserve between stops in the event weather cuts him off from some of the planned alternates along the way. If reason tells the pilot that the expected weather may bring an early darkness, he will want to estimate and add to his night-flying fuel reserve.

Pilot fatigue also requires consideration. A tired pilot is filled with poor judgment and wrong decisions. Each pilot should know his own endurance and get himself on the ground for a rest before he starts seeing dual instrumentation on the panel. He should be aware that a weather briefing of less than "heaven with a fence around it" can shorten his fair-weather endurance. Flying through turbulence, beneath low clouds, or between showers, or poor visibility all make a pilot's job a little harder. It's natural that he will tire more quickly and should counter this with planned and needed rest stops. Remember, even the wind aloft can exact a toll. A pilot's jargon of "fighting a head wind" and "riding a tail wind" are more than colorful metaphors. They are expressions of real-life expended effort.

In sum, as you obtain your weather briefing, write down the weather factors that are expected to prevail and evaluate the actual effects that they will present in the cockpit. Once done, you can feel confident that your preflight weather preparation has been on target.

In Review

- The first step toward understanding a weather briefing is to stay in tune with changing weather patterns on a day-to-day basis.
- Three charts provide a pilot with a quick overall weather picture: the surface analysis chart, the weather depiction chart, and the radar summary chart.
- The surface analysis chart shows the location of fronts.
- The weather depiction chart outlines areas of en route IFR and marginal VFR conditions.
- The radar summary chart displays areas of precip.
- Give the weather briefer three statements as you ask for the preflight briefing:
    1. Whether flying IFR or VFR
    2. Destination, route, alternate airports, and desired altitude
    3. Proposed time off, time en route, and any intermediate stops
- Reduce the weather briefer's verbal information to a written weather log for an invaluable en route reference.
- Compare the weather briefing to your personal weather minimums to initially determine a *go* or a *no-go* weather picture.
- Relate the weather briefer's information to the effects it will have on cruising altitude, route, and refueling stops.
- Factors that influence a pilot's choice of altitude include:

    Winds aloft
    Freezing level
    Cloud levels
    Turbulence
    Aircraft capability
    Transponder equipment

Available oxygen
Passenger comfort

- Factors that influence a pilot's choice of route include:

    Areas of precip
    Areas of low visibility
    Areas of turbulence
    Areas of low clouds
    Position of fronts
    Availability of alternate airports
    Areas of heavy traffic
    Unlit areas at night

- The weather briefing affects your preplanned refueling stops in two ways: aircraft capability and pilot fatigue.

Preflight Aids

WEATHER LOG

DEPARTURE AIRPORT

	CURRENT WEATHER	FORECAST WEATHER
CLOUD BASES & COVERAGE		
VISABILITY		
PRECIP./ FOG		
SURFACE WIND		

ENROUTE WEATHER

CURRENT WEATHER AT _____	
CLOUD BASES & COVERAGE	
VISABILITY	
PRECIP./ FOG	
SURFACE WIND	
NOTAMS / PIREPS :	

ROUTE FORECAST	CONDITION/TIME/LOCATION
WINDS ALOFT	
LOW CLOUDS	
LOW VISIBILITY	
PRECIP./ FOG	
ICING	
TURBULENCE	

DESTINATION & ALTERNATE

	CURRENT AT DESTINATION	DESTINATION FORECAST AT ETA	ALTERNATE FORECAST AT ETA
CLOUD BASES & COVERAGE			
VISABILITY			
PRECIP./ FOG			
SURFACE WIND			

A written log often provides an invaluable inflight reference.

CHAPTER 10: **Preflighting the Destination Airport**

IT WAS A SLEEPY SUMMER DAY. The pilot's third go-around brought the local airport crew to their feet. They left their makeshift chairs of cased Aeroshell and gray steel toolboxes and moved as a body through the open doorway of the wooden hangar out into the warm afternoon sunshine. All eyes focused upward and outward along the turf airstrip's departure path. Echoes of the revved 200 HP Lycoming still bounced through the beech- and pine-covered knolls surrounding the small Appalachian airport.

The band of airport lookers caught sight of the plane banking around to downwind for another try. On this one, the pilot tucked his pattern in tight and closed his throttle with a determination that the folks on the ground could sense. The airport gang then scurried over to the vantage point by the sock to watch the only Saturday afternoon excitement the sleepy mountain airfield could mount—a caught-by-surprise pilot trying to squeeze his long-runway airplane into a short-strip airport.

Finally, the sleek retractable sank through base leg like a wounded U-boat, turned short final, did a little Stuka dive over the fence, and arrived with a whump that would do any carrier pilot proud. Hard braking took divots out of the turf as the bouncing plane sped past the wind sock and swung to a lurching stop in front of the line shack. I thought the passenger looked a

Preflighting the Destination Airport

bit tight around the mouth and eyes as he climbed to the ground. But the pilot was a picture of confidence as he bent down to count the wheels, ordered the tanks topped, and went off in search of a taxi. "Rented aircraft," I thought to myself.

I had a flight on and could not stay to watch the pilot try to track down Charlie, the town's only cabbie. When next I saw the pilot it was mid-afternoon two days later when he returned to the airport with his passenger in tow, but not following too closely, I noticed. The pilot loaded the baggage, and the passenger climbed aboard. The pilot went through the starting sequence. Gerald, the airport owner, saw what was happening but he did not make his move until the plane's prop was turning. (He likes a dramatic touch; he delays his help to the last minute.)

Gerald stepped in front of the churning engine and made a chopping motion ordering the pilot to shut down. Gerry explained: "It is possible to land in a field that you can't take off from."

The airport owner and pilot then reviewed the plane's performance charts. On arrival the plane needed 1,960 feet; a snug fit for the 2,400-foot mountain strip. With the topped-off tanks and the mid-afternoon sun, however, the handbook called for a 2,480-foot takeoff. But if the pilot waited for the next morning's cool weather and de-fueled to a third of capacity, the needed distance dropped to 2,140, and the pilot could refuel at the large airport twenty minutes down the valley. Gerald was about to suggest that the pilot and passenger enjoy another night in the mountains, but then saw that the passenger was already mounting a search for Charlie and his cab.

That departure, of course, could have been researched before the flight ever left home. All the information was at hand: field elevation, expected temperatures, runway surface, and the aircraft takeoff chart that would have advised against topping off the tanks on arrival.

Much of the preflight information is available in the *Airport/Facility Directory* on display at all FSS, many fixed base operators, and available from the government on a single issue or subscription basis. A copy belongs in every preflight planning kit. With it, a pilot can relate the destination airport to the facts and figures of his plane's performance tables, the weather briefing, navigational and communication needs, and his own level of skill.

Preflight planning information within the directory includes:

Runway data. Each runway heading is represented in the form of a runway number and can easily be compared to the forecast surface wind for crosswind and headwind components. Additionally, the width of each runway is stated—important to know when a pilot is considering whether to tackle a formidable crosswind. Runway length and surface appear for comparison with takeoff and landing charts and pilot skill, along with a description of obstacles that might get in the way of an arrival or departure. The field elevation and traffic pattern altitude are stated, as are any nonstandard right-hand pattern turns.

Communication and Nav Aids. All communication facilities that the airport offers are listed along with the frequencies. Any limitations such as part-time operation are noted, as is special use airspace such as a TCA or TRSA.

Radar services, VORs, and ADFs are described, along with any unusual range restrictions.

General airport information. The airport's three-letter identifier used for flight plan filing is displayed. The availability of fuel and the type of repair services offered are stated, as is airport lighting. If the lighting system makes use of pilot-activated runway lights, operating instructions are given. Any unusual flight hazards such as skydiving, a one-way strip, construction in progress, or closed runways are noted.

FLORIDA

ORLANDO FSS (ORL) on Orlando Executive JACKSONVILLE
123.65 122.65 122.2 122.1R 118.7 (305) 894-0861 H-4G, L-19

ORLANDO
MAGUIRE (X30) 7 W GMT−5(−4DT) 28°31'50"N 81°32'26"W JACKSONVILLE
130 S4 FUEL 100LL
RWY 18-36: 2430X140 (TURF) LIRL
RWY 18: Trees. RWY 36: Thld dsplcd 325'. Road.
AIRPORT REMARKS: Attended 1300-0100Z‡. After hours call (305) 656-3191. Gliders rgt tfc when rwy 36 in use.
COMMUNICATIONS: UNICOM 122.8
ORLANDO FSS (ORL) LC 894-0861

ORLANDO EXECUTIVE (ORL) 2.6 E GMT−5(−4DT) 28°32'43"N 81°19'59"W JACKSONVILLE
113 B S4 FUEL 100, JET A H-4G, L-19C
RWY 07-25: H5998X150 (ASPH) S-45, D-65, DT-115 HIRL IAP
RWY 07: SSALR. Trees. RWY 25: REIL. VASI(V4L)—GA 3.0° TCH 45.9'. Trees. Rgt tfc.
RWY 13-31: H4618X100 (ASPH) HIRL
RWY 13: VASI(V4L)—GA 3.0° TCH 27.59'. Trees.
RWY 31: VASI(V4L)—GA 3.0° TCH 27.86'. Tree. Rgt tfc.
AIRPORT REMARKS: Attended 1200-0300Z‡. Arpt closed to acft over 100,000 lbs. Fee for all charter, travel clubs and certain revenue producing acft. Brightly lgtd bridge highway located approximately 1/2 mi. S. of arpt. Could give false indication of being rwy on apch to Rwys 07 & 31 during low ceiling or poor visibility. VFR acft arriving/departing Orlando Executive Arpt exercise caution due to turbo-jet acft transiting Orlando Executive Arpt traffic area 2000'/above on approach to Orlando Intl Arpt 5.6 miles south.
COMMUNICATIONS: ATIS 127.25 UNICOM 122.95
ORLANDO FSS (ORL) on fld 123.65 122.65 122.2 122.1R 112.2T (305) 894-0861
® ORLANDO APP CON 124.8 (4000' above 337°-179°) 120.15 (4000' above 180°-336°)
121.1 (below 4000' 250°-070°) 119.4 (below 4000' 071°-249°) 125.55
EXECUTIVE TOWER 118.7 (1200-0400Z‡) GND CON/CLNC DEL 121.7
® ORLANDO DEP CON 124.8 (4000' above 337°-179°) 120.15 (4000' above 180°-336°)
121.1 (below 4000' 250°-070°) 119.4 (below 4000' 071°-249°)
STAGE III SVC ctc APP CON
RADIO AIDS TO NAVIGATION:
(H) VORTAC 112.2 ORL Chan 59 28°32'33"N 81°20'07"W at fld. 110/00
VOR unusable 050°-060° beyond 15 NM below 5000'.
HERNY NDB (LOM) 221 OR 28°30'24"N 81°26'03"W 070° 5.4 NM to fld
ILS 109.9 I-ORL Rwy 07. LOM HERNY NDB
ASR
COMM/NAVAID REMARKS: Freq 118.7 remoted to FSS for AAS when tower closed.

ORLANDO INTL (MCO) 6.1 SE GMT−5(−4DT) 28°25'54"N 81°19'29"W JACKSONVILLE
96 B S2 FUEL 80, 100, JET A LRA CFR Index D H-4G, L-19C
RWY 18R-36L: H12004X300 (CONC) S-100, D-200, DT-400 HIRL IAP
RWY 18R: MALSR. VASI(V4L). Rgt tfc.
RWY 18L-36R: H12004X200 (ASPH-GRVD) S-165, D-200, DT-400 HIRL CL
RWY 18L: VASI(V6L)—Upper GA 3.25° TCH 89.7'. Lower GA 3.0° TCH 52.4'. Thld dsplcd 989'. Pole.
RWY 36R: ALSF2. TDZ. VASI(V4L)—GA 3.0° TCH 52'. Rgt tfc.
AIRPORT REMARKS: Attended continuously. LLWAS. CAUTION—Birds and deer on and in vicinity of arpt. Rwy 18L-36R slippery when wet. Flight Notification Service (ADCUS) available.
COMMUNICATIONS: ATIS 121.25 UNICOM 122.95
ORLANDO FSS (ORL) on Orlando Executive NOTAM FILE MCO
® APP/DEP CON 124.8 (337°-179°) 120.15 (180°-336°)
TOWER 124.3 GND CON 121.85 CLNC DEL 134.7
STAGE III SVC ctc APP CON
RADIO AIDS TO NAVIGATION:
(H) VORTAC 112.2 ORL Chan 59 28°32'33"N 81°20'07"W 173° 5.7 NM to fld. 110/00
ILS 110.7 I-OJP Rwy 36R
ILS 111.9 I-TFE Rwy 18R
ASR
COMM/NAVAIDS REMARKS: Orlando Ramp 121.35 (1200-0400Z‡) provides ramp advisory service East Terminal Ramp Area.

Much of the necessary preflight data concerning a destination airport is available in the Airport/Facility Directory.

The value of a preflight study of the airport's Instrument Approach Procedure Charts is obvious to an IFR flight. But the VFR pilot can also draw good preflight information from them. Probably the handiest item on the charts lies in the aerodrome sketch. This provides a scale drawing of the airport's active runways and taxiways, handy indeed for positive in-flight identification of a runway. Handy, too, as a map to assist you in that toughest navigation problem of all—how to taxi your way from the ramp area to the active runway. The minimum sector altitudes of the approach's overhead view give an arriving pilot a guaranteed safe height that misses any obstacle within twenty-five miles of the approach's primary nav aid. The depicted final approach course assists the VFR pilot in avoiding high-speed instrument arrivals.

A pilot's preflight review of the destination airport should include a study of his sectional chart. If the flight is to an unfamiliar destination, a preflight selection of prominent landmarks helps a pilot zero in on the airport. He should also pick a prominent landmark from which to enter the traffic pattern. All too often, the arriving pilot flies a beeline to the airport from his direction of travel. If he does this without regard to the traffic flow in the pattern, he is often forced into a hazardous hairpin turn simply to avoid head-on traffic on the downwind leg. Reason out the traffic flow beforehand by simply correlating forecast wind to the runways available. Then select a prominent landmark several miles from the pattern, which allows a forty-five-degree entry to the downwind leg.

The sectional chart also gives the pilot a preview of the traffic he might encounter on arrival. A second or third airport within a few miles of his destination field is good cause for raising the guard, as is any nearby military airfield or metropolitan airport with scheduled airline traffic.

Preflighting the destination airport moves up in priority when

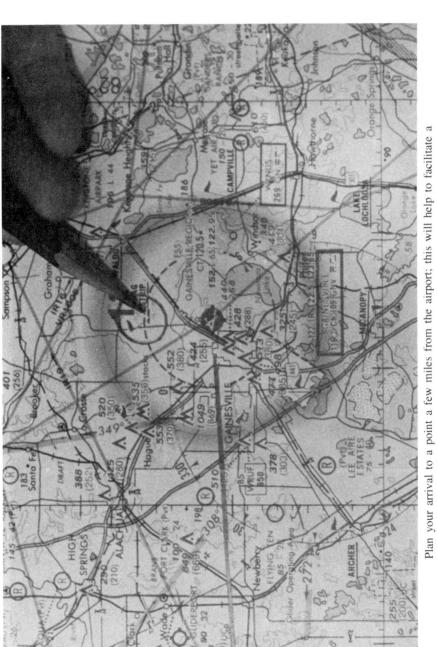

Plan your arrival to a point a few miles from the airport; this will help to facilitate a correct pattern entry.

the cross-country flight is a long one; the arriving pilot is often tired. The cockpit work load normally increases in the terminal area. And any amount of this work that the pilot disposes of before the flight begins eases the pressures of arrival. The peace of mind that a pilot feels as he nears journey's end is directly proportional to the effort he expended in preflighting the destination airport.

In Review

- It is possible to land in a field that you can't take off from.
- You may want to consider de-fueling, ferrying passengers one at a time to a nearby larger airport, or waiting for a cooler, drier runway.
- Nearly all the information you need to preflight the destination airport is at hand. Gather the facts from the *Airport/Facility Directory,* Instrument Approach Charts, and the Sectional Chart. Correlate these items with the weather briefing, aircraft performance, and level of pilot skill.
- Preflighting the destination airport moves up in priority when the cross-country flight is a long one and the arriving pilot is tired.

Preflight Aids

AIRPORT DATA FOR _____

FIELD ELEVATION _____ ft.   PATTERN ALTITUDE _____ ft.
RIGHT HAND PATTERN RUNWAYS _____ _____ _____
RUNWAYS WITH OBSTACLES   _____ _____ _____

ACTIVE RUNWAYS

R/W Number & Length	R/W Number & Length
_____ _____ ft.	_____ _____ ft.
_____ _____ ft.	_____ _____ ft.
_____ _____ ft.	_____ _____ ft.

RUNWAY LIGHTING

Full Time _____ Partial Hrs. _____ To _____
By Prior Request _____ Pilot Controlled _____ MHz

Maintainence & Fuel Available _____
Recommended FBO _____

Airport Remarks / NOTAMS :

Runway Layout :

Extract the essential airport data from the Airport/Facility Directory and enter it into a destination log for easy reference on arrival.

CHAPTER 11: **Compiling the Radio Frequency Log**

WE WHO SHARED THE SKY with Sam described him as a pilot who was intensely likable, had a profound sense of survival, and was almost totally devoid of basic pilot skills—particularly navigation. Many times Sam would stop by the airport lobby on his way to the tie-down to ask if any student was about ready to taxi out. Sam would ask the student to wait a minute so he could follow the student to the active runway.

Sam was no better in the air. Compass and chart bore no apparent relationship to him. And the "to" and "from" business with the VOR made no sense whatever.

The way Sam flew, covering his Florida sales territory by plane was really taking the slow way. An automobile, with friendly gas station attendants to point the way, would have been faster. But Sam loved flying . . . hanging there between cloud and ground, watching his sales territory pass in review. Yet he knew, even before takeoff, that he would get lost before his wheels again landed him back on his home airport.

You might think it impossible to get lost and stay lost over peninsular Florida. After all, morning and afternoon all a pilot has to do is point his nose at the sun, wait for the coast to show up, then turn either left or right to find an identifiable airport within ten minutes.

Sam knew he could do it; he didn't worry too much. He held

great faith in the FSS specialists and ATC controllers. They would find him and return him safely enough.

Before each flight he carefully prepared a written log of the radio frequencies that connected him to his navigational system—assistance from the ground. The system worked well enough. The controllers came to know him, what with his friendly manner and his distinctive accent acquired between Lebanon and Brooklyn. Sam sensed that the controllers enjoyed his company despite their sometimes feigned annoyance. It was the time it took to get unlost that bothered Sam; all that time it took to work radioed instructions for cross-fixes and fly identifying turns was robbing his plane's efficiency. ("Piper eight three Tango, turn left heading three three zero, and . . . left, Sam . . . *left* . . . *LEFT!*") Sam thought there had to be a better way.

One noon while having a chicken-salad sandwich with a controller on his route, Sam learned the answer. The controller told him of an avionic just becoming available to light planes—a transponder. With one installed in his plane, Sam would forever be in sight of a radar scope, and the controllers could simply guide him along the route as they passed him from sector to sector. Sam would even be able to dial in his own code and he never again would need to fly an identifying turn. Sam was ecstatic. Finally someone had invented a device that made an airplane useful. Sam bought a transponder that day.

Month after month the three of them flew his routes—Sam, his log of frequencies, and his transponder. His flights were unerringly true. He would just tune in the sector frequency, tell the controller where he wanted to go, and simply follow the controller's guiding instructions. An Eastern Airlines captain could fly no straighter path.

Then one afternoon while beep-beep-beeping his transponder up the south Florida coastline, Sam saw a black storm ahead. A small airfield lay just ahead of his nose so he dove for it and quickly landed. He stopped by the side of the runway, shut down, and ate an apple while the storm took twenty minutes to pass over him. Then he took off, climbed to the altitude that started his transponder winking a reply, pulled the ATC frequency from his radio log, and checked in with the controller. "*Sam,*" the anxious controller shouted, "where have you been?"

No matter what the pilot's experience level, a radio frequency log prepared during preflight makes life aloft easier and safer. A radio frequency log is simply a listing of the needed flight frequencies. It is true that some of the frequencies are printed on the navigation chart. But finding the frequency one-handed while flying with the other is another matter. The chart handles with all the ease and common sense of unfolding a bargain-basement chaise lounge.

Normally, all the frequencies that a pilot needs are contained in the small carry-aboard-sized *Airport/Facility Directory*. But the typesetter that prints that book prints Bible verses on pin heads. Trying to thumb out a frequency in the thick little volume while the fresh-air vent flaps the pages is an exercise in temper control. A simple radio frequency log is the logical answer.

To compile your log, just look along the course line you have drawn on your sectional chart and think ahead to the facilities you will encounter on your flight. Then jot down the frequencies in a form similar to the sample radio log that follows. Here are some of the facilities that you may encounter along your way

AUTOMATIC TERMINAL INFORMATION SERVICE (ATIS)

ATIS is a continuous broadcast of takeoff and landing information normally encountered at heavy-traffic airports. Its purpose is simple: to relieve radio congestion as the ground or tower controller gives taxiing or landing instructions to departing or arriving aircraft. It is basically a rebroadcast of the airport's hourly weather sequence report and contains in order:

1. The time of the weather sequence report.
2. Cloud layer height and coverage.
3. Visibility and the condition responsible for any low visibility.
4. Temperature.
5. Wind direction and velocity. (Direction is converted to magnetic.)
6. Kollsman setting.
7. The instrument approach in use.
8. The runway in use.
9. The approach control frequency in use.
10. The phonetic alphabetical code word that identifies the current ATIS broadcast.

The departing or arriving pilot is expected to monitor the ATIS frequency prior to calling the ground or tower controller. He should advise the controller that he has obtained the latest information, by using the phonetic identifier. The controller may then abbreviate his instructions to the pilot, omitting those items in the broadcast. (The pilot's phrase "have the numbers" is not an adequate notice that he has received the ATIS broadcast. The phrase means he has the runway number, wind, and altimeter setting only. The controller must then repeat the other ATIS information.)

GROUND CONTROL

After listening to the ATIS broadcast, a departing pilot normally contacts ground control. The ground controller has jurisdiction over the taxiways, and the departing pilot must be in radio contact before entering a taxiway. (A taxiway is denoted by a continuous yellow centerline. The edge may be marked with dual yellow lines.)

Before opening the microphone, the departing pilot should mentally review the five-part initial radio contact:

1. Identify the station. ("Orlando/Exec. ground control.")
2. Identify aircraft. ("Mooney two two three five Bravo.")
3. Give location on the airport. ("Piper ramp . . .")
4. State request. ("Taxi for takeoff . . .")
5. Advise latest ATIS received. ("Information Tango.")

The controller's taxi clearance to the runway allows the pilot to cross any intervening runways (excluding the active runway itself) in order to reach the active runway, unless the clearance contains instructions to "hold short" of a runway or intersection. Often the ground controller specifies routing to the active. This route then becomes mandatory.

Getting from the ramp to the active runway can be a confusing bit of navigation at an unfamiliar airport. Do not hesitate to conclude your initial call-up with the statement "Need directions." The controller will give you guidance to the runway. (Controllers do not mind lending this service in the least; it is part of the job. What they *do* mind, however, is an unfamiliar pilot taxiing on his own to an incorrect spot that blocks the flow of traffic.)

The ground controller's clearance limit is to the "holding lines," normally one hundred feet short of the runway entrance. Holding lines consist of two solid and two dashed lines, perpendicular to the taxiway centerline. The taxiing pilot cannot roll past these lines without further clearance for the area beyond falls under the jurisdiction of the tower controller.

CONTROL TOWER

The tower controller's area of responsibility includes the active runway and the airport traffic area. The airport traffic area is a cylinder of airspace five miles in radius and 3,000 feet tall that sits over any airport with an operating control tower. Its confines are not depicted on the sectional chart. In preflight planning keep in mind that an airport traffic area overlies any blue airport symbol, then estimate a five-mile radius around that symbol.

The departing pilot must contact the tower for takeoff clearance before he crosses the holding line at the runway entrance. The pilot needs only to advise the tower that he is "ready for takeoff." The tower then may issue one or all of the following instructions:

"Hold short" (Don't roll onto the runway.)

"Position and hold" (Roll to the runway centerline, but do not start the takeoff.)

"Cleared for takeoff" (The tower expects the pilot to be prepared and to move without delay.)

The tower may modify these three basic clearances somewhat. Most modifications are clearly understandable, such as "move closer and hold short" or "position and hold with room for one more behind you." But the modified "cleared for immediate takeoff" bears some discussion. When the controller uses the term "immediate takeoff," he is expressing an option to the pilot: If the pilot has the experience to execute an immediate takeoff, he is cleared to do so, but if he lacks the experience he should advise the tower that he will "hold short." Strange things can happen during an attempted immediate takeoff by an inexperienced pilot; decide before you taxi to the runway whether you can handle tower instructions for an "immediate takeoff" without becoming flustered. Once in the air the pilot is required to monitor the tower frequency for additional instructions until he either clears the airport traffic area or is issued a frequency change by the tower controller.

The controller's permission must be obtained also by any en route pilot wishing to fly through an airport traffic area. Preferably his initial call should occur fifteen miles from the airport and should state his present position and his request to fly through the airport traffic area. This gives the controller time to evaluate his own arriving and departing traffic so that he can issue specific clearance instructions to this pilot. Pilots overflying a controlled airport in excess of 3,000 AGL often erroneously call the tower for permission. In actual fact they are operating outside the jurisdiction of the airport traffic area. Instead these pilots should contact approach control for traffic advisories.

In preplanning figure that an arriving pilot needs to contact the tower far enough out to allow the controller planning time—about twelve to fifteen miles for a plane running over 150 knots; around eight to ten miles for a slower one. A five-part initial contact is appropriate, similar to a departing pilot's taxi clearance transmission:

1. Identify the station. ("Orlando/Exec. tower.")
2. Identify the aircraft. ("Beechcraft two two five one Delta.")
3. Give position. ("Twelve northeast . . .")
4. State request. ("Landing Orlando/Exec.")
5. Advise latest ATIS received. ("With information Kilo.")

Once in the traffic pattern, a pilot should refrain from any unexpected maneuvering and let common logic prevail. The controller can anticipate, for example, shallow S-turns for spacing. He cannot second guess, however, an unannounced 360-degree turn. (This maneuver usually scatters traffic like a blast of bird-shot whistling through a covey of quail.)

As the landing pilot is rolling out, the controller normally gives the pilot the ground frequency to contact *after* leaving the runway. A common mistake here is for the pilot to change frequency *before* he leaves the runway and crosses the holding line.

DEPARTURE/APPROACH CONTROL

This is really one facility. Call it "departure control" if leaving or "approach control" if arriving. This facility normally provides VFR radar service. A departing pilot who desires radar service (and pilots are urged to participate) should make his wishes known to ground control at the conclusion of his request to taxi (". . . request VFR radar service, northbound"). The tower will then tell the pilot when to change frequency to ATC—normally shortly after takeoff.

Arriving pilots should prepare to contact the approach control twenty-five miles out from the airport. The initial contact should simply state the plane's position from the airport, altitude, transponder code (if transponder-equipped), destination, a request for VFR radar service, and the latest ATIS received.

Depending on the radar service advertised in the *Airport/Facility Directory,* the pilot may expect to receive basic VFR radar service, Stage II service, or Stage III service.

Basic radar service means that the controller issues traffic advisories and, if requested by the pilot, vectoring to the traffic pattern. These basic services are offered on a "time permitting" basis. Approach control provides runway and wind information if the pilot has not received an ATIS broadcast. They also specify when or where you are to contact the tower.

Stage II service provides sequencing as well as the basic radar services. Pilot participation is not mandatory, but the controller assumes the radioing pilot wants it unless he specifically states "negative Stage II service." In that event the tower controller issues the landing sequence when the pilot enters the traffic pattern. The pilot accepting full Stage II service, however, holds the advantage of having his landing traffic pointed out to him several miles from the airport. Once the pilot advises the radar controller that he has the plane he is to follow in sight, the pilot just trails the plane into the pattern.

Stage III offers full radar service to the arriving VFR pilot: traffic advisories, vectoring, altitude assignments, and aircraft separation. Again, if the pilot advises that he does not want Stage III, the service reverts back to basic traffic advisories. The participating full service Stage III pilot receives vectors and altitudes that place him behind the traffic he is to follow. Once positioned behind the preceding airplane and upon reporting that he has the aircraft in sight, the pilot maintains visual separation to touchdown.

VFR radar service in the terminal area offers an incalculable safety factor to a pilot busy with his own arrival chores. There are three things he must remember, however. First, in VFR conditions midair avoidance is the prime responsibility of the pilot. He must view radar service as just another tool to use in carrying out that responsibility. Second, the radar controller cannot see IFR conditions. If a heading or altitude assignment will plunge the pilot into instrument conditions, he is obligated to advise the controller and obtain an alternate heading and altitude. And finally, the pilot should realize that VFR radar services often create misunderstandings between himself and the controller. This is no one's fault in particular. It is just that VFR radar procedures are not as formalized as IFR procedures. When the pilot has a question in his mind about the controller's intent, he should simply ask.

ATC CENTER

Most of the busier airways have radar service available between terminal areas through Air Route Traffic Control Centers. The Center's primary responsibility is to provide separation of IFR flights. These facilities, however, provide radar assistance to VFR pilots on a "time permitting" basis. Their service is invaluable when visibility drops or the rain begins to cluster.

FLIGHT SERVICE STATION

In addition to bearing the prime responsibility for preflight briefings, FSSs also:

Provide en route VFR communications
Accept and close flight plans
Assist lost VFR aircraft
Provide airport advisory service

Let's look in more detail at these FSS services.

En route VFR communications. When en route cross-country, I make it a practice to use the first of these services. I contact each FSS that my path crosses. On this call I obtain the current altimeter setting (an FAR requirement every hundred miles) and an update on the weather ahead. If necessary, I advise of any change in ETA that exceeds my flight plan by twenty minutes.

An initial call-up is in three parts:

1. Identify the station you are calling ("Orlando radio . . ."). Adjacent FSSs often share a common frequency. Both hear you if you are high enough and they need to know whom you want to hear from.

2. Identify yourself (". . . Cessna four six two two Hotel . . ."). As in all contacts with ground facilities, give your aircraft make and full number on initial contact. On subsequent

calls to the facility, abbreviate your call sign to the aircraft make and last three characters of the aircraft identification number.

3. Advise the frequency over which you expect a reply (". . . listening one two two point six five"). The FSS specialist monitors several frequencies that all come through the same speaker. While there is a light that flickers the triggered frequency, the specialist may be looking at a weather report or flight plan when it winks. If he has not heard your desired frequency, he must go to the trouble of keying all of his frequencies to reach you.

Once you have completed your three-part initial call-up, wait until the specialist responds before you proceed with your message. He may be busy with another pilot on another frequency. For this reason, delay for at least one minute any repeat initial call-up.

Accepting and closing flight plans. The FSS accepts and closes flight plans. The purpose of a VFR flight plan is to assure the pilot of a timely search and effort rescue, should he become overdue. According to the National Search and Rescue Plan (an interagency effort), "The life expectancy of an injured survivor decreases as much as 80 percent during the first 24 hours." An Air Force review of 325 search missions over a twenty-three-month period reports: "Time works against the downed pilot without a flight plan, since 36 hours normally pass before family concern initiates an alert." The arithmetic here is simple; file a flight plan with the FSS to insure a timely search and rescue should you go down.

Assist lost aircraft. When compiling your preflight radio frequency log consider the services of a DF (Direction Finding) steer. This DF equipment has long served to locate lost pilots and guide them to better weather or airports. The FSS procedure is simple, requires only a two-way radio, and is almost conversational in its execution. FSS specialists invite pilots to give them a call for a practice demonstration. Should the pilot need the service for real, he has in a sense "been there."

Airport advisory service. FSSs are sometimes located at an airport either without an operating control tower, or with a part-time closed tower. When the FSS is the sole agency on the field it renders airport advisory service. This service gives arriving and departing pilots a reliable source of runway information, but it cannot exercise any traffic control. The pilot can reach this service on either a standard 123.6 MHz or the part-time tower's frequency; check the *Airport/Facility Directory.*

FLIGHT WATCH

The services of En Route Flight Advisory (Flight Watch) will update your pre-flight weather briefing. Flight Watch is a network of highly trained weather briefers stretched across the USA. They are located at only four dozen FSSs across the country, but by virtue of their many remote transceivers, the specialists offer a continuous blanket of coverage for planes at or above 5,000 feet, with reasonable contact as low as 2,000 feet.

All in the Flight Watch network share a common 122.0 frequency, and specialist/pilot communications are very informal. Their weather advisories come primarily from the in-flight reports of pilots just through the area that you are asking about. So expect a nuts-and-bolts weather briefing pertinent to your route, altitude, and the type of equipment you are flying.

Expect also to be questioned about the weather surrounding your plane so they can pass it along to other pilots following you. Before you call Flight Watch, take time to make your own weather observation. Be as accurate as you can. Relate the visibility, for example, to the farthest landmark you can see and compare this to the mileage on your sectional chart. Estimate cloud bases by comparing the angle from your known altitude to the ground against the angle upward to the clouds. Cloud shadows on the ground give a good indication of "scattered" or "broken," and bodies of water plainly indicate surface wind. Use the standard terms to describe conditions such as turbulence and write down your facts and figures. Then give Flight Watch your well organized pilot report with specifics that are meaningful to the pilots that follow. (A report that says ". . . sorta bumpy and I can't see very far . . ." really doesn't give another pilot much to go on.)

UNICOM

A unicom facility is simply a radio station operated by a fixed base operator. Although it has no official ATC status and cannot offer aircraft separation or landing sequence, a pilot may wish to include the unicom frequency in his preflight planning. The facility is of some limited use in obtaining the active runway and wind conditions at uncontrolled airports, and the information may or may not be accurate.

By reporting his inbound position five miles from the airport, reporting downwind and listening for other pilots' position reports, a pilot can gain *minimal* traffic information. The main hazard of using this method is that a pilot who is intent on making a unicom report (of little real importance) may fail to maintain a vigilant search for traffic (of vital importance). Make use of unicom, certainly. But do not let it replace your own logic for landing information or your own eyes for traffic separations.

EMERGENCY

Any preflight preparation should include the possibility of an emergency. Nearly all FSS and ATC facilities monitor 121.5, the emergency frequency. This frequency is there for any pilot to use, and he should include it in his radio frequency log. If a pilot is already in touch with a facility on the normal frequency when an urgency arises, he is better advised to seek help on that normal frequency. This simply precludes the loss of communication with a twist of the frequency selector.

Most pilots commit the same common error when making calls for assistance; they delay calls until the situation has grown into a full-blown emergency. By this stage, pilot options are so limited that the information supplied from the ground is of little value.

When should you call for assistance? Apply one rule: Call the moment you feel a genuine concern about your fuel supply, the weather, a malfunctioning plane, or your exact position.

I believe that pilots who delay their calls do so for two reasons. First, the pilot is flustered or anxious and cannot decide whom to call or take the time to search out a frequency. Second, the pilot hesitates to do anything as dramatic as "declare an emergency" or holler "Mayday . . . Mayday . . . Mayday" until he is certain he has a truly desperate situation. By that time outside help is of little use. I suggest pilots prepare a log of frequencies, and if they cannot decide whom to call, then just ring 121.5. It is not necessary to make a dramatic declaration or holler heroic phrases. When you feel a concern for your safety just open the mike and say, "I need some help!" You will get it.

General aviation safety has come a long way from the pioneering days of the 1920s, 1930s, and 1940s. I think the achievement in safety is the result of three factors: better weather forecasting, better informed and trained pilots, and better radio communications with all the services that those facilities imply. A simple, well-organized radio frequency log puts those services at your immediate fingertips.

My friend Sam who was a hopeless navigator gave up flying

a few years ago at the insistence of his loving family. I think it was a wise decision. Yet whenever I'm flying and open a frequency, I still half expect to hear Sam and some controller helping him navigate his territory. ("Piper eight three Tango, descend now to two thou . . . Sam, dip your nose till the big hand's on zero and the little hand says two. . . .")

In Review

- A simple radio log makes a pilot's life aloft a lot easier; the FAA publications are not easy to handle in the air.
- All frequencies are listed in the *Airport/Facility Directory*.
- To compile a log, look along your course line across the sectional chart and anticipate the facilities you will encounter.
- The pilot's phrase "have the numbers" is not adequate notice that he has received ATIS.
- A departing pilot must contact ground control before he enters a designated taxiway.
- If you do not know the airport, ask the ground controller for directions.
- An airport traffic area is five miles in radius, 3,000 feet high, and overlies any airport with an operating control tower.
- VFR radar service in the terminal area offers an incalculable safety factor.
- It remains the pilot's responsibility to avoid traffic and IFR conditions.
- If you do not understand the intent of a radar controller's instructions—ask.
- Air Route Traffic Control Centers offer invaluable assistance to VFR flights when the visibility drops or the rains gather.
- File a flight plan with FSS to insure a timely search and rescue, should you go down.
- FSSs invite pilots to ask for a practice DF steer.
- Carefully prepare any weather observation that you forward to Flight Watch.
- Do not become so intent on calling unicom that you fail to keep a vigilant traffic watch.
- Call for assistance the moment you feel a genuine concern about your fuel supply, the weather, a malfunctioning airplane, or doubts about your exact position.

Preflight Aids

RADIO FREQUENCY LOG FOR FLIGHT

FROM _____ TO _____

COMMUNICATION FREQUENCIES

STATIONS:				
ATIS				
Ground Control				
Tower				
Depart APCH Control				
ATC Center				
Flight Serv. Station				
Flight Watch				
UNICOM				
Emergency				

NAVIGATION FREQUENCIES

STATIONS:				
VOR				
ADF				

A simple radio log is easier to read aloft than charts or directories. To compile your log, just look along your course line and anticipate the facilities you will encounter.

CHAPTER 12: **Preplanning the Navigation**

DOUGLAS CORRIGAN was a light-hearted teenage mechanic who soloed just before he went to work for the Ryan Aircraft Company. It was 1927 and the fledgling company, occupying an old brick two-story warehouse on the run-down side streets of San Diego, had just landed a contract to build a small monoplane, the *Spirit of St. Louis*.

Corrigan was a cheerful dreamer with a ready friendship for anyone in reach. He had been orphaned as a boy with a younger sister and brother to support; his young life already knew hard work and determination. But his determination had paid off. He kept his family intact, and he achieved his dream, which was to fly. And here he was, working on the airplane that would fly Charles Lindbergh from New York to Paris. He had reason to be cheerful.

Imagine the impact the youthful Lindbergh had on the younger pilot Corrigan when he visited the shop to work on his plane. There was only one larger dream possible. Someday Corrigan hoped to fly the Atlantic. He hovered nearby any time Lindbergh came to the factory. (Many photos taken of Lindbergh working on his plane include Corrigan, standing five feet five inches tall in the background, staring with a grin about as wide as his skinny shoulders.) Corrigan hung on every word he could catch. He listened as Lindbergh discussed navigation, true course and variation, winds aloft, and drift angles. If dead reckoning navigation was the magic carpet across the ocean, Corrigan would learn how to use it.

Corrigan left the Ryan factory shortly after Lindbergh made his famous solo flight to Paris. Corrigan left behind the wrenches and screwdrivers and took up the gypsy life of the barnstormer. For the next ten years he flew a dilapidated biplane in and out of any small town pasture that had paying customers and a friendly sheriff. When the flights between pastures were long, he would tuck away a Texaco road map and navigate by dead reckoning. The dream persisted.

Then in 1938 Corrigan thought it was time. He needed a better craft than his patched-up biplane and looked around for a suitable ocean-hopping dreadnought. He had $325 in cash to spend. Corrigan ended up with a ten-year-old Curtiss Robin of dubious maintenance history. The plane's rotting fabric didn't keep out much sunlight, and the wooden ribs held together only because the termites held hands. Even in its day the Robin was not meant for long-range flying; a Sunday spin around town for the plane was what Curtiss had in mind.

Corrigan tried to fix all the plane's ailments. He fitted it with extra homemade gas tanks (they leaked), and he replaced the original worn-out, wheezing OX-5 engine with a tenth-hand gasping Whirlwind that was only half dead. He was ready for the Atlantic.

Corrigan flew the rattletrap nonstop from his home in California to New York City (astounding), practicing his dead reckoning for the ocean flight that lay ahead. He then applied to the U.S. Department of Commerce for permission to fly the Atlantic. The Department's inspector took one look at the crate and gave his flat refusal. Corrigan then asked permission to fly back to California nonstop with a vague explanation about some sort of westbound record. The inspector said that was not his responsibility and left the scene.

So Douglas Corrigan lifted from New York's Roosevelt Field early on the morning of July 17, 1938. He took off east into the wind, climbed east into the sun, and continued flying east, making himself into an American legend. Corrigan later claimed to government officials and the press that he got turned around in the clouds and simply flew the wrong way. He even explained, poker-faced, how he read the compass backwards and thus flew east instead of west. That incredible pilot error was supposed to have lasted the twenty-eight hours it took the tired plane and engine to clank and belch its way across the Atlantic.

What Corrigan *did* do was pay close attention to his dead reckoning. The blown wave tops of the gray North Atlantic provided accurate wind direction and velocity, the compass a course line, and the torturous 80 MPH airspeed was only too clear on the panel before him. Each hour he carefully marked his attained position along his drawn course line; if the method worked for Lindbergh, it would work for him.

Corrigan finally crossed the Irish coast very near his intended landfall, found the airport he was looking for, and landed. Before onlookers reached his plane, Corrigan quickly cleared his charts away, stepped from the airplane, and asked: "Where am I?" He managed an appropriate look of surprise when told he had miraculously chanced upon Dublin's fair city.

Then he spun the tale that would become aviation's greatest hoax. Said he: "I took off from New York to fly nonstop to California . . . haze hid the ground and I got turned around in the clouds . . . read the compass backwards, I guess . . . anyway, I flew a day and descended . . . sure didn't look like California . . . used two hundred ninety gallons of gas . . . that's sixty-nine dollars and twenty cents." And he stuck to his story. Average pilot skill, basic dead reckoning, and a jalopy airplane came together to give a young aviator his slice of immortality. His name, "Wrong Way" Corrigan, remains a household word to this day.

Just as in Corrigan's day, dead reckoning is a vital part of preflight calculations. It is sometimes said that dead reckoning navigation is the lost art of flying; lost somewhere among those pale blue lines of the VOR airways. It is true that many pilots place their entire navigational trust in the vertical needle of the omni head. But there is a basic flaw in blindly following a wavering needle cross-country. The needle bears little resemblance to what is really happening to the plane and where it is flying. The needle does little to help the pilot feel the force of the wind playing against the strength and course of the plane. Further, it cannot help us visualize the flight path we are on as an angle to the section lines, and it cannot help us picture the terrain that will pass beneath the plane. As a result any needle or VOR follower is often left wondering just where he is.

To achieve the feel of the cross-country flight, the pilot needs to employ dead reckoning and its companion, pilotage (chart and landmarks). But even dead reckoning has its objects of deception—computer and plotter. A pilot who blindly takes for granted the results of turned aluminum discs or imposed plastic rulers is as much in the dark as the blind needle follower. I have seen pilots transpose the values of ground speed and distance on their computer and faithfully carry the resulting impossible time en route to the cockpit. I have observed others who have carefully plotted the reciprocal of their desired course. And I have seen even more who have laid their plotter WAC side up and accepted a mileage that flies them two states beyond their destination. These were silly mistakes, of course. They are the kinds of errors that throw confusion into the cockpit; the kind of confusion that prompts a pilot to call ATC and ask for real: "Where am I?"

By contrast the dead reckoning pilot who learns to estimate rather than calculate gains the feel for his plane and sky, because in order to estimate he must *visualize*. For this pilot the wind becomes a visible force either pressing back against the left side of the nose cowl or lending a push to the right side of the vertical stabilizer. The pilot sees the course as a path at angles with a section line rather than as a small number on the humped back of a plastic plotter. The miles suddenly become visualized as real rolling landscape from departure to destination. Ground speed at the same time becomes a percentage he can compare with his mile-a-minute automobile. That pilot can tell you his ETA almost to the second. In short, the estimator holds all the advantages over the calculator. The charts following this chapter are presented with this in mind—to help you hone your preflight estimating skills.

To be a skillful estimator, the pilot needs to know the factors to weigh. Let's return to our student-pilot days and review the language of dead reckoning navigation. Here are some of the terms we encounter:

True Course. This is the course you draw on the sectional chart during preflight preparation. True course is the direction of travel measured from the zero line of longitude and is related against the geographic north pole. Before you lay your plotter to that line, estimate the angle across a line of longitude.

True Heading. The term "heading" in navigating a plane means that a wind correction has been applied to the true course to compensate for drift. As you preplan your navigation draw an arrow across your course line from the direction of winds aloft. There is no question, then, whether to subtract from or add to true course for your true heading.

The amount of wind correction depends upon the wind velocity, the angle it blows across the true course, and the plane's

Preplanning the Navigation 225

airspeed. The wind correction angle may be calculated on a computer, or it may be accurately estimated with the table that follows this chapter.

Magnetic Heading. In addition to the wind correction, in preflight charting a pilot needs to modify his course to allow for

MAGNETIC VARIATION

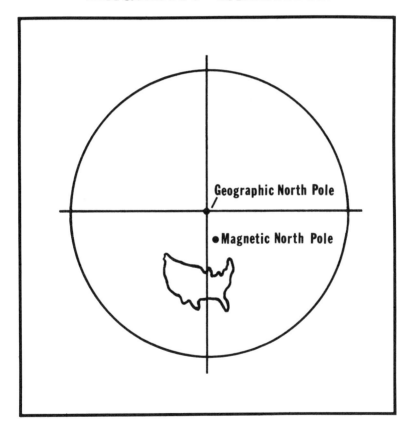

A correction for magnetic variation simply allows for the fact that *true* north (to which our lines of longitude lead) lies several hundred miles from *magnetic* north (to which our compass points).

magnetic variation. Magnetic variation is simply a navigational term that recognizes the fact that magnetic north (to which our compass points) is several hundred miles from geographic north (to which our course line is related). The amount of correction that a pilot needs varies from location to location and is shown by the *isogonic* line on the sectional chart that lies closest to his route. (Add the degrees of correction shown on the isogonic line if the variation is shown as ''west.'' Subtract if the variation is ''east.'')

Compass Heading. Once the pilot is in the cockpit, he must allow for the error that his individual compass contains. (It is difficult to keep them error-free.) The needed correction is called compass deviation and is found on a card just beneath the compass.

Pilots often ask why all this conversion of directions is necessary. The answer is simple. *Wind* must be reported from true north, the only value that is consistent over the large area of an air mass, yet the compass that we follow will point only to magnetic north. The conversion steps are needed to compare wind to flight path just as it is necessary to compare apples to only apples. We convert everything to magnetic.

Ground Speed. The same wind that causes drift affects a plane's ground speed. The degree that the wind affects the ground speed depends on the wind's velocity, angle of head wind or tail wind, and the plane's airspeed. Ground speed may be calculated or accurately estimated with the table that follows.

Distance. For distance during my preflight calculations I use the VOR roses to estimate mileage across my sectional chart. These roses are about ten miles in radius. Using this scale, I just use thumb and forefinger to pace off the mileage, as that is accurate enough for legs under two hundred miles.

Time En Route. Time en route is dependent on ground speed

Preplanning the Navigation

and distance. Use the table at chapter's end to estimate the time required for each leg of the flight.

Fuel Needed. To obtain the fuel needed, divide your engine's fuel flow into the estimated time en route. Apply your desired reserve to preplan refueling stops.

VOR Radials. The VOR stations are aligned with magnetic north for the area in which they lie. The courses marked on the airways are magnetic and need no correction for variation.

Midpoint Actual Time of Arrival. Mark the leg's midpoint actual time of arrival on your sectional chart. Then note the time of takeoff and the time you pass over your midpoint. You will then have a revised time of arrival at leg's end, and you may want to reevaluate your fuel remaining.

In Review

- Dead reckoning and pilotage are the primary means of navigation for the VFR pilot. They depend on visual references—landmarks and visualized wind.
- A pilot who learns to estimate rather than calculate often has a more realistic picture of his cross-country flight.
- True course is the line on the chart.
- True heading is really true course corrected for wind drift.
- Magnetic heading is true heading corrected for magnetic variation.
- Compass heading is the magnetic heading corrected for instrument error.
- Ground speed should be thought of as a percentage of your mile-a-minute (60 MPH) automobile on the highway.
- Use the ten-mile radius of the VOR rose to pace off distance.
- Reestimate your ETA and fuel situation at the midpoint of your cross-country flight.

Preflight Aids

ESTIMATING WIND CORRECTION ANGLE

WIND VELOCITY AND ANGLE		AIRCRAFT AIRSPEED AND WIND CORRECTION ANGLE		
		90–120	125–155	160–200
10–20 KTS.	Within 30° of Nose or Tail	3°	2°	1°
	30° to 60° off Nose or Tail	6°	4°	3°
	60° to 90° off Nose or Tail	8°	6°	5°
25–35 KTS.	Within 30° of Nose or Tail	4°	3°	2°
	30° to 60° off Nose or Tail	12°	8°	6°
	60° to 90° off Nose or Tail	16°	12°	10°
40–50 KTS.	Within 30° of Nose or Tail	5°	4°	3°
	30° to 60° off Nose or Tail	18°	12°	9°
	60° to 90° off Nose or Tail	24°	18°	15°

ESTIMATING GROUND SPEED

(Percentage of windspeed to add
or subtract from tailwind or headwind)

Wind within 30° of T.C.	± 100% of Windspeed
Wind 30° to 60° off T.C.	± 60% of Windspeed
Wind 60° to 90° off T.C.	± 30% of Windspeed

ESTIMATING TIME AND DISTANCE

At a G.S. of 90	Each 100 miles takes 1:05
At a G.S. of 120	Each 100 miles takes :50
At a G.S. of 150	Each 100 miles takes :40
At a G.S. of 180	Each 100 miles takes :33

NAVIGATION PLANNING LOG

_____ TO _____

LEG FROM – TO	TRUE COURSE	TRUE HDNG. (wind)	MAG. HDNG. (var.)	GROUND SPEED	DISTANCE OF LEG	TIME ENRTE.	FUEL NEEDED	VOR STATION / RADIAL	MID-POINT ATA
									..
									..
									..
									..
									..
									..

AIRCRAFT _____

TRUE AIRSPEED _____

FUEL FLOW _____ / HR.

FUEL ENDR. ____ : ____ HR.

FUEL STOPS

_____
_____
_____

CLOSE FLIGHT PLAN BY :

☐ RADIO WHILE AIRBORNE
☐ PHONE AFTER LANDING
☐ OTHER _____

CHAPTER 13: **Complying with FARs**

FIVE REGULATIONS in particular concern themselves with the pilot's environment. We are apt to think of a pilot's environment only in terms of weather. Total environment, however, includes runway conditions, radio communications, traffic, and special use airspace. As with the FARs concerning the pilot and the airplane, a thorough initial reading of the selected regulations is recommended. Let the capsulized versions that follow serve as a quick preflight reminder.

FAR 91.5 PREFLIGHT ACTION

Checking the weather reports and forecasts, calculating the fuel requirements, and planning alternate airports for cross-country flights are not only good operating procedure; FAR 91.5 requires these steps. The regulation requires the pilot to include these specific items in his preflight planning for "a flight not in the vicinity of the airport." So far as I know, the "vicinity" of this FAR is not defined. But you may want to use my yardstick: I consider myself beyond the "vicinity" when I look back and cannot tell just where the airport lies. My "vicinity" thus changes with the weather or light—scattered showers, visibility, height of cloud base, or amount of daylight. I feel this definition is in keeping with the intent of the FAR and serves my purpose well.

The regulation in addition requires a pilot to determine the runway lengths at departure and destination and compare them to the performance capability of his airplane.

FAR 91.87 OPERATION AT AIRPORTS WITH OPERATING CONTROL TOWERS

Airports that have control towers are marked on the sectional chart as blue aerodrome symbols and are surrounded by an airport traffic area. This piece of special use airspace is not defined on the chart, but it is a standard cylinder of airspace five miles in radius and 3,000 feet high. If your route of flight takes you through an airport traffic area, make certain that your radio contains the tower frequency. FAR 91.85 requires communication with the tower during any flight through a control tower area.

FAR 91.90 TERMINAL CONTROL AREA

Pilots planning to operate in a Terminal Control Area (TCA) are required to have at least a private pilot certificate. The aircraft must have a radio with the appropriate ATC frequency, a VOR receiver, and a transponder with altitude-reporting capability.

FAR 91.95 RESTRICTED AND PROHIBITED AREAS

Check your intended route for restricted or prohibited areas. Authorization can often be obtained before the flight in order to operate through a restricted area. FSS can supply the name and phone number of the controlling agency and often has a direct line to them.

The existence of a restricted area close to your route is fair warning that you may encounter fast-moving traffic. FSS can verify the type of operations normally conducted in and around the area.

FAR 91.109 VFR CRUISING ALTITUDES

When cruising above 3,000 AGL, fly odd thousands plus 500 feet when traveling a *magnetic* course of 0 to 179 degrees (3,500, 5,500, 7,500). A magnetic course of 180 to 359 degrees requires even thousands plus 500 feet (4,500, 6,500, 8,500).

CHAPTER 14: Evaluating the Environment for IFR Operations

THERE ARE SEVERAL PREFLIGHT EVALUATIONS of the environment that have special significance to the departing IFR pilot. A recent instrument flight that I made between Orlando and Tampa illustrates this point. The flight was my instrument student's last training flight before he was scheduled to take his IFR flight test. We went through the preflight planning together, taking special care to request an altitude that would keep us in the clouds for his final student flight. Conditions were perfect for the occasion; IFR all the way with good VFR 2,000 feet below our planned altitude for the first half of the trip, and with clear skies just thirty miles south of our airway. At the destination, conditions were 600 broken with two miles and light rain, with a fairly heavy flow of traffic. The student was going to get a good parting shot at the real stuff before he became a full-fledged instrument pilot.

Our preflight briefing finished, we took off, followed radar vectors to join Victor 152, and settled down to the flight. Then midway down the southwestward airway an unusual emergency hit us. It was the sort of situation that is just not contemplated by IFR regulations or procedures. In fact, most of the textbooks seem to hesitate talking about the problem. Halfway down the airway our entire group of avionics lost power and quit. Not just communications, but everything including the VOR. So with the terrain blanked by cloud, we could not navigate the airway, or stand any chance of making an approach at the destination.

The student saw the "Off" flags fly, but held his existing heading and altitude (that earned him an A+) while he took a moment to ask the next step:

Student: "What do we do?"
Me: "Can we stay IFR?"
Student: "Nope . . . gotta get VFR."
Me: "How about going down . . . it's VFR under us?"
Student: "Nope. There's two layers of IFR under us, too."
Me: "OK . . . how about the VFR south?"
Student: "Ummm . . ." He looked at the chart that showed two airways between us and the clear skies.
Me: "How can we miss those airways?"
Student: "Ummm . . . aha, we go down five hundred feet and cross 'em on a VFR altitude."
Me: "Let's go."

So he eased us off the IFR altitude, turned to one eight zero, and neither of us said a word for the fifteen minutes that we followed the gauges to the sunshine. Five minutes after that we found an airport, landed, and phoned ATC to quell their curiosity. Our preflight search for VFR conditions along our route had paid off.

Here are several items of preflight planning and preparation that bear a special relationship to the IFR pilot and his environment:

THE PROXIMITY OF VFR CONDITIONS

The pilot planning to fly a light plane through instrument conditions should plan his route so that VFR is within reasonable reach. An IFR pilot will probably never experience that total loss of avionics. There are, however, many other causes that call for a fast transition to VFR conditions. Consider, for example, the single-engine IFR pilot who sees that the engine's oil pressure is dropping and the temperature is rising. That pilot wants to get it on the ground in a hurry. If the nearest instrument airport is still thirty minutes away, he will welcome vectors to nearby VFR and a VFR airport. The same yearning for quick VFR is felt by the pilot who experiences VOR antenna problems, iced up wings, or any other condition that prevents safe continued IFR. The VFR conditions can be off to the side or immediately beneath. But the pilot of a light plane in IFR conditions should, if at all possible, have VFR conditions within easy reach.

TURBULENCE

To a VFR pilot, moderate turbulence means discomfort. To the light plane IFR pilot, however, turbulence takes on additional meaning. Put together extensive IFR, a hand-flown airplane, and moderate turbulence, and you end up with a fatigued pilot. If that fatigued pilot doesn't have a deep reserve of instrument experience to fall back on, he is apt to start making a string of bad decisions and pilot errors.

INSTRUMENT LANDING COMPONENTS

A vital preflight item concerns the navigational components of the published instrument approach procedures at the destination, alternate, and (in case a turn-back is required) departure airports. The IFR pilot needs to study the approach procedure charts for these airports and compare the equipment required with the equipment he has on board. If the destination or alternate airport only offers an NDB approach, the pilot had better have a working ADF on board.

An approach chart that describes the procedure as VOR/DME *requires* distance measuring equipment. (An approach described in the margin identifier as simply VOR, however, does not require distance measuring equipment, even though a DME arc might be shown. The DME in this case is at the pilot's option.)

An ILS approach requires the following components:

Localizer
Glide slope
Outer marker
Middle marker

Several substitutions are allowed, however, if the aircraft is missing some of the needed equipment, or if the ground components are inoperative:

1. If the pilot lacks a glide slope, he may execute the approach as a localizer approach. His preflight, however, had better confirm that the VOR receiver of that rented aircraft *has* localizer capability; many do not.

2. The approach controller's radar identification of the outer and middle markers will substitute for the lack of a marker beacon in the airplane. The pilot's preflight, however, had better make certain that the airport has radar services available.

3. At some airports (shown on the approach chart) DME is an allowable substitute for a middle or outer marker.

4. If a compass locater is positioned at the marker, ADF may substitute for the marker beacon.

RADIO COMMUNICATIONS FAILURE

The IFR pilot should take a few minutes during his preflight preparations to review the standard procedures to follow, in the event he loses voice communications with ATC. Radio communications failure is not an uncommon occurrence. In the event it happens, the controller will expect the pilot to follow specific procedures. As soon as the pilot realizes he has lost voice communication, he should adjust his transponder to code 7700 (emergency) for a period of one minute. Following his one-minute alert, he should then change to transponder code 7600 for a period of fifteen minutes. After that time has lapsed he should repeat the sequence every sixteen minutes. The pilot should keep in mind, however, that he may not be in radar coverage.

If the two-way radio communication failure occurs in VFR conditions, or if VFR conditions are encountered after the failure, the standard operating procedure requires the pilot to remain VFR, land as soon as practicable, and close his flight plan by phone.

If the aircraft's radio fails in IFR conditions, and VFR conditions are not subsequently encountered, the controller expects the pilot to follow specific procedures concerning the aircraft's route and altitude.

ROUTE

The pilot should continue his flight through the instrument conditions according to these steps:

1. By the route assigned in the last ATC clearance that he received

2. If being radar vectored, by the direct route from the point of radio failure to the fix, route, or airway specified in the vector clearance

3. In the absence of an assigned route, by the route that the controller has advised may be expected in a furthur clearance

4. In the absence of either an "assigned" or "expected" clearance, by the route he filed in his flight plan

ALTITUDE

The "radio-out" pilot should continue his flight through the IFR conditions at the *highest* of these altitudes:

1. The altitude assigned in the last clearance received
2. The minimum altitude specified on the chart as the minimum en route altitude (MEA) for the route segment being flown
3. The altitude ATC has advised may be expected in a further clearance

As an example of this procedure in action, let's assume that a pilot experiences radio failure just after receiving an assigned altitude of 7,000 feet, while flying on a route segment with a 6,000 MEA. He would fly the higher of the two altitudes—7,000 feet. If the next route segment had an MEA of 8,000 feet, he would climb to 8,000 feet (the higher altitude) after passing the fix defining the segment. He would then follow this "highest altitude" procedure to his clearance limit or to VFR conditions.

If this pilot had been told to expect a 9,000-foot assignment at a specified fix along the airway, then 9,000 feet would enter the "highest altitude" procedure *after* the pilot reached that specified fix. In addition to the last assigned altitude, expected altitude, and minimum en route altitude, the pilot must, of course, comply with any "minimum crossing altitude" requirement along the airway.

If instrument conditions persist to the destination, ATC expects the pilot to conclude his flight according to these procedures:

Leaving the clearance limit. If the pilot has received an "expected further clearance" (EFC) time, he should leave the clearance limit at that time and proceed to the holding pattern depicted on the chart for the initial approach point.

If an EFC time was not received prior to the radio failure, the pilot is expected to proceed directly from the clearance limit to the initial approach fix. If the chart shows more than one initial approach fix, the pilot may select the one he wishes to use, because ATC has protected his airspace at each of them.

Descending for the instrument approach. The "radio-out" pilot should begin his descent from the en route altitude upon reaching the approach fix, but not before the ETA that he filed on his flight plan. If he arrives at the approach fix before his ETA, he should fly a holding pattern at the final fix on the "procedure turn" side. He can then begin his descent in the holding pattern at the moment of his ETA, knowing that ATC has then cleared all traffic beneath him, between the final approach fix and the runway.

If the controller's radio fails, the pilot should delay changing to another frequency for one minute. This allows ATC's backup system time to reestablish radio contact. If radio contact is not reestablished the pilot should then try another frequency. If the next sector frequency is known (or determined from the pilot's radio log) he should try to reestablish communications through that facility. Failing in this effort, the pilot is expected to contact FSS, which is in direct interphone contact with ATC and can relay instructions quickly.

It is in the pilot's interest to comply with the standard procedures for coping with radio communications failure. ATC service and aircraft separation will be provided on the assumption that the pilot knows the procedures and will follow them.

PREFERRED ROUTES

Preferred routes are "one-way" airways that connect large airports with heavy traffic flows. They are established to increase the efficiency and capacity of these busy airways, and they will dictate the routing given in the departing IFR pilot's clearance.

A pilot's preflight should determine if any preferred routes lie along his proposed flight. All preferred routes are listed in the *Airport/Facility Directory,* alphabetically by the terminal area they serve. If a preferred airway does lie along the pilot's proposed flight, he should file for a routing that accommodates these "one-way" airways. If he does not, and tries to fly the wrong way on a one-way airway, ATC will arbitrarily change the pilot's expected route. Then during clearance delivery the pilot will undoubtedly receive a complicated amended clearance. For many pilots this is a highly distracting occurrence.

MINIMUM CROSSING ALTITUDES

Instrument pilots of low performance aircraft should make a preflight examination of their route depicted on the low-altitude en route chart. They should look for minimum crossing altitudes (MCA) that might demand more from their plane's climb performance than it is capable of delivering.

The pilot of a low-performance fully loaded aircraft cruising at 8,000 feet, for example, might expect his plane to deliver a 250 FPM rate of climb. If an 11,000 MCA lies only ten minutes ahead, he can expect a possible delay at that point while he gains the altitude.

A preflight review of the chart would have divulged this potential problem. The forewarned pilot could then alert the controller to the possible delay, and get a higher altitude assignment well in advance of the MCA.

STANDARD INSTRUMENT DEPARTURES (SIDS)

At many heavy-traffic airports, ATC has a standard departure procedure. They have established these procedures to simplify clearance delivery, speed the flow of departing traffic, and help the pilot avoid inbound traffic or significant obstacles. Most SIDs involve a complex exit from the airport area. For this reason the departures route, altitudes, and required airspeeds are published in both text and graphics. These SID descriptions are published in booklet form, include about a hundred airports for which SIDs are established, and are updated every two months.

The departing pilot's clearance delivery assumes that he has the published SID before him and will comply with its demands. If the pilot does *not* have the SID booklet, he is expected to advise ATC of the fact when he first files his flight plan. In all probability he will then be read the text of the SID during his clearance delivery. This calls for a significant journalistic effort on the pilot's part as he then tries to copy the complex clearance.

This annoying distraction at the outset of an instrument flight can be avoided by a preflight review of the SID booklet. Each airport that has an established SID is listed alphabetically in the booklet's index. The forewarned pilot can then have the appropriate SID chart before him during departure and exit the area with little trouble.

STANDARD TERMINAL ARRIVALS (STARS)

The same hundred or so airports that have Standard Instrument Departures also have Standard Terminal Arrivals. A STAR is simply a SID coming the other way. STARs are established procedures designed to accommodate the flow of inbound traffic. Again, the pilot will prevent the need to copy a lengthy and complex clearance by simply having the appropriate STAR chart before him. As was the case with SIDs, these arrival procedures are bound in booklet form for the IFR pilot's preflight review.

TRANSITION AREA

Some airports that have a published instrument approach available do not offer the protection of a *control zone*. Control zones are established around most airports that offer an instrument approach. The zones prohibit uncontrolled VFR operations during those times of instrument conditions. The arriving IFR pilot, then, is assured that all aircraft around the airport are under the control of ATC; his approach path through the clouds is protected.

The existence of a control zone, however, is dependent on a local weather observation to determine if IFR conditions prevail. Yet some airports with published instrument approaches do not have facilities to observe and report the weather. These airports, then, have no control zones. They have instead a transition area; an area that means an instrument approach is available but without the protection of a control zone. (These transition areas are easy to spot on the sectional chart during preflight. They are shown as a *magenta* circle around the airport symbol, but without the broken blue outline of the control zone.)

To the arriving IFR pilot coming down the final approach, a transition area means that the possibility of a midair collision exists. It is not uncommon, for example, for the IFR pilot to break out of the clouds and find an unannounced VFR plane turning from base to final just yards ahead.

IFR pilots should determine before the flight whether their instrument approach involves a transition area. Then, forewarned, they will know to be extra cautious as they break out of the clouds on final approach.

TOWER EN ROUTE CONTROL (TEC)

In many parts of the country an IFR pilot can fly instruments from one airport to another without ever leaving approach control. The ATC service that makes this possible is called Tower En Route Control, and it allows IFR flight beneath the en route structure. The TEC concept pairs together airports that lie within the range of one another's radar coverage. If a pilot can accept a low-altitude IFR flight the TEC will save him considerable filing time, as he does not enter the *Center's* airspace. (Most TECs limit the pilot's choice of cruising altitudes to those below 5,000 AGL.)

The airports that are paired together by a TEC are listed in the *Airport/Facility Directory*. The available routes are spelled out, along with the highest altitude available. Although *all* airports are not listed, the IFR pilot may use the described route to plan an approach to an airport that lies close to the named major airport. Flight plans should be filed in the normal manner with FSS, with the notation TEC entered in the flight plan's remarks section.

These several additional items of preflight planning and research add substantial safety and efficiency to IFR operations. The following combined chapter review and preflight aids will serve as a convenient checklist of items for the instrument pilot to consider as he evaluates the environment for IFR operations.

IN REVIEW

The following is a checklist of several additional items of preflight planning that are of special interest to the departing instrument pilot.

- The Proximity of VFR Conditions: Light-plane IFR operations should have VFR conditions within reasonable reach.
- Turbulence: Mix together moderate turbulence and a handflown airplane with extensive IFR conditions, and you end up with a fatigued pilot who is susceptible to pilot error.
- Instrument Landing Components: The IFR pilot needs to compare his plane's navigational equipment with the equipment required by the approach at his destination.

    An ILS approach requires:

    Localizer
    Glide slope
    Outer marker
    Middle marker

    If the pilot's plane lacks a glide slope receiver, he may execute a nonprecision localizer approach. Radar, DME, or a compass locater may substitute for the outer and middle marker.

- Radio Communications Failure: Review text as a preflight reminder of "radio-out" procedures.
- Preferred Routes: Check the *Airport/Facility Directory* to see if a preferred route (one-way airway) pertains to the airports along your route.
- Minimum Crossing Altitudes: Research the low-altitude en route chart for any MCAs that might exceed the climb performance of your plane at cruise altitude.
- Standard Instrument Departures: Check the SID booklet to see if your departure airport requires a standard instrument departure procedure.

- Standard Terminal Arrivals: Check the STAR booklet for the existence of a Standard Terminal Arrival at your destination airport.
- Transition Area: If your destination is a small airport, check the sectional chart for the lack of control zone protection.
- Tower En Route Control: If your trip is to a nearby terminal area, check *Airport/Facility Directory* for the availability of a low-altitude TEC filing.

CONCLUSION: A Pilot's Responsibility

THE GOAL OF EVERY PILOT should be professionalism. Amateur flying is a dangerous and often frustrating sport; safety and satisfaction lie in professionalism. Being a professional pilot has nothing to do with flying for hire. Rather, professionalism has to do with the pilot's attitude and sense of responsibility toward flight; he must have a willingness to learn the limits of his plane and bring his own skill up to meet those limits. He must be willing, also, to evaluate his own judgments and misjudgments in order to virtually eliminate pilot error. Further, he must be willing to acquire a total knowledge of flight with which to anticipate the unexpected.

A pilot owes professionalism to himself and his passengers. Taking a plane aloft is not a child's game. The stakes are high. And any pilot who turns his mind away from this reality is only kidding himself.

For a moment, think about the tradition of professionalism that we have inherited. The flying that means so much to us today came from the efforts, trials, and even terrors of the pilots who preceded us. Charles Lindbergh was a true professional. He paid incredible attention to the smallest detail; he had an absolute devotion to perfection, defying the criticism and warnings of the less professional. Amelia Earhart, another pioneer, taught us a good deal about professionalism before her untimely death in the Pacific area.

In addition we can draw professional strength from thousands of less famous pilots who flew before us. I think of the faceless bush pilot skimming the featureless tundra under a rapidly lowering overcast, desperate for a firm landable patch of survival; and I think of the nameless pilot of the night mail, numbed with fatigue, peering over the rim of his cockpit through the sting of snow for the beacon that winks "rest." When we think of these pilots, we should sit a little straighter and try to make our flying worthy of the legacy.

We have another—perhaps greater—obligation, an obligation to those pilots who will follow us and draw strength from our achievement. Just as sure as Champion makes little ceramic spark plugs, some Sunday afternoon in the year 2195 a weekend pilot is going to be tooling over the arctic circle at 60,000 feet in his Mooney Mark XXXXIV, and when he punches Oslo's coordinates into his Narco Lasergator he will think of you and me pushing our hand-flown machines through the thick weather-laden band of mile-high air. I believe he will think of you and me and sit a little straighter and fly a little better to be worthy of *our* achievements . . . and to provide the example for pilots who are to follow him.

The brotherhood of pilots (fraternity of airmen, or whatever you call it) is there to join, but our flying must be good enough to keep the chain of professionalism unbroken. I hope this book will be useful in further developing your own professional goals.

INDEX

Index

Abort/stop distance, 117-18
Accidents, causes of
  alcohol, 72
  antagonism, 60
  false sense of security, 10
  get-home-itis, 62
  inadequate preflight planning, 1
  "it won't happen to me" syndrome, 61
  low clouds, 24
  night flying, 42
  poor pilot condition, 56, 190
  precipitation, 36
  propeller strikes, 136, 149
  reduced visibility, 11
  rushing, 58
  strong crosswinds, 40
  trim control, lack of, 155
  uncontrolled airports, 19
  violation of FARs, 66
Aircraft preflight inspection
  checklist items, 136-46, 158-59
  pilot attitude toward, 135
  required effort and, 135
Airport/Facility Directory
  airport lighting, 196
  FAA district offices, 94
  preferred routes, 242
  radio frequencies, 196
  radio test checkpoints, 167
  radio test (VOT) signals, 167
  runway information, 196
  tower en route control (TEC), 245

Alternate airports and flying in marginal conditions, 26, 28, 34, 36
Altitude
  communication failure and, 239-41
  low flying, 29-31
  minimum crossing altitudes (MCA), 242
  over the top flying, 30-33
  regulations concerning, 163, 164, 234
  selection of, 20, 182-87
Automatic Terminal Information Service, 205-206

Basic truth of flying, 1
Before-takeoff checks, 150-56
Best-angle climb speed, 117
Best-rate climb speed, 117
Biennial flight review, 68-69
Brotherhood of pilots, 249

Checklist
  airframe items, 136-46, 159-60
  before takeoff, 150-56, 160
  familiarity with, 146
  pilot condition, 65
  starting engine, 146-49, 158-59
Control tower, 208-10
  regulations pertaining to, 233
Corrigan, Douglas ("Wrong Way"), 220-23
Crosswinds
  estimating crosswind component, 41
  night flying and, 44

[253]

Crosswinds (continued)
  personal experience with, 41
  personal weather minimum for, 40
Cruise performance
  fuel management and, 121
  fuel mixture and, 119
  in practice flight, 80
  proper power settings for, 118
  proper trimming and, 122

Dead reckoning navigation
  in estimating the flight, 28, 224
  factors to weigh for, 224-27
Departure/approach control, 210-12
  radar assistance and, 18, 32, 36
  and terminal areas (TCA), 233
Disorientation and reduced visibility, 13

Earhart, Amelia, 248
Emergency services
  direction finding (DF) steers, 213-14
  during low-altitude flying, 30
  ATC radar assistance, 18, 32, 36
  when to request, 215-16

Federal Aviation Regulations (FARs)
  aircraft documents, 163
  airframe inspections, 164
  airworthiness, 164
  biennial flight review, 68
  compensation for the pilot, 71
  control towers, 233
  cruising altitudes, 234
  false sense of security and, 10
  fuel requirements, 161-62
  high-performance aircraft and, 67
  liability and civil action, 66
  liquor and drugs, 72
  logbook entries, 67
  medical certificates, 67
  medical deficiency, 68
  oxygen and, 164
  pilot defiance of, 60
  for pilots, 67
  preflight action, 232
  radio navigation equipment, 163
  recent flight experience and, 68
  responsibility and authority of pilot, 71
  restricted areas, 233
  terminal control areas, 233
  transponders, 163, 164, 233
Field elevation
  and landing distance, 123
  and takeoff performance, 115-16
Flap position
  before takeoff, 115
  landing performance and, 122
  during low-flying navigation, 29
  takeoff performance and, 114
Flight Service Station (FSS), 212-14
  inflight communications and, 31, 35
  weather briefings from, 178-82
Flight Watch, 214-15
  inflight weather advisories from, 15, 28
Fuel management
  before takeoff, 154-55
  while cruising, 118-22
  low-altitude flying and, 29
  preflight inspections, 138-39, 139-40, 153
  and priming engine, 148
  refueling stops, 189-90
  regulations concerning, 161-62

Getting lost, causes of
  DF steers, 213-14
  basic pilot tool, 1
  fatigue, effects of, on, 190
  stress, effects of, on, 16
Gross weight
  landing distance and, 123
  takeoff performance and, 115
Ground control, 206-207
  radio equipment check and, 165

Headwind
  fatigue and, 190
  landing distance with, 123
  takeoff performance with, 115
Heading indicator, 153

Index

Icing
  of carburetor, 30, 148
  effects of, 185
  intensities of, 45, 184-85
Ignition switch
  and airframe inspection, 136
  runup, 156
"It won't happen to me" syndrome, 61-62

Judgment
  basic pilot tool, 1
  fatigue, effects of on, 190
  stress, effects of on, 16

Knowledge
  Advisory Circulars, 95
  cornerstone of competence, 1
  preflight readiness, 94
  testing and evaluation of, 94

Landing gear handle, 138
Landing minimums, 45
Landing performance
  factors influencing, 122-25
  during practice flight, 81
Lindbergh, Anne Morrow, 7
Lindbergh, Charles A.
  preflight attitude of, 134
  professionalism of, 248
  and Ryan Aircraft Company, 220
Low clouds
  hazards of, 24
  and personal weather minimums, 25
  precautions to take in, 26
  selecting a route to avoid, 189

Magnetic compass
  before-takeoff check of, 152
  and compass deviation, 226
  and magnetic variation, 225-26
Master switch, ammeter malfunction of, 148-49, 154
Midair collisions
  basic defenses against, 18-24
  and low-altitude training routes, 27
  low clouds and, 24
  reduced visibility and, 13
  responsibility of pilot for, 211
Mixture control
  before takeoff, 154
  cruise range and, 119
  high elevation takeoffs and, 121
  and hot starts, 148

Navigation
  and current position, 28, 31, 35
  by dead reckoning, 16
  by pilotage, 16
  by radio navigation, 16
  simultaneous navigation, 17

Obstacles
  and landing performance, 124
  low-altitude flying and, 29
  low clouds and, 25
  and takeoff performance, 116-18

Passengers
  comfort of, 36-37, 187
  losing friends, 35
  putting to work, 18
  requirements for carrying, 70
Personal experience, importance of gaining, in
  crosswinds, 41
  low clouds, 25
  night flying, 24
  precipitation, 34
  reduced visibility, 14
  short field operations, 118
Personal weather minimums
  elements of weather, 11
  extent of precipitation and, 33
  IFR considerations, 45
  low clouds and, 24
  reduced visibility and, 11
  strong crosswinds and, 40
  turbulence and, 36
  written record of, 46

Pilot errors
  inadequate planning and, 1
  personal potential for, 53
  and preflight planning, 5
  stalls and, 88
Power settings
  common misconceptions, 118-19
  manufacturer's recommendations, 119
Precipitation
  night flying and, 43
  personal weather minimums for, 33
  when selecting a route, 187-88
Professionalism, 248
  obligation of pilot, 248-49
  related to total awareness, 4
Propeller
  before-takeoff check of, 155
  and power settings, 118-19
  and preflight inspection, 142
  safety, 136, 149

Radio equipment
  backup items, 168
  and engine start, 147
  failure of during flight, 238-41
  frequency log, 205
  preflight checks of, 165-69, 173
  radio installation, 168
  in terminal control areas, 233
Reduced visibility
  hazards of, 11
  radar services during, 18
  safeguards during, 14
  when selecting a route, 189
Routing
  alternate airports, 26
  to avoid heavy traffic, 20
  following roads, 23
  and night flying, 43
  preferred routes, 241-42
  and radio failure during flight, 239
  standard instrument arrival and, 243
  standard instrument departure and, 242-43
Runway surface conditions
  Airport/Facility Directory and, 196

  at alternate airports, 28
  landing distance and, 123
  regulations on, 233
  strong crosswinds and effect on, 40
  and takeoff performance, 114

Saint Exupéry, 175
Seat belts
  before-takeoff check of, 156
  and turbulence, 39
Skill
  as basic pilot tool, 1
  in crosswinds, 40
  frequency of flight and, 75
  practice flights and, 76
  preflight readiness and, 75
Stalls
  causes of, 89
  low-altitude flying and, 29
  recovery procedure for, 91
  and weight and balance, 125
Stress
  getting lost and, 17
  personal capacity for, 57

Takeoff minimums, 44
Takeoff performance
  factors affecting, 114-18
  and leaning, 121
  practice drills and, 77
Temperature
  of engine, 153
  freezing level, 184
  landing performance and, 123
  overheating of starter, 149
  takeoff performance and, 116
Thunderstorms
  and acceptable terrain coverage, 33
  selecting a route to avoid, 188
  and turbulence, 38
Total awareness, 4
Total preflight planning, 4
  aircraft capability and, 113
  aircraft condition and, 135
  and environment, 177, 196, 205, 233, 235

Index

pilot capability and, 10, 75, 94
pilot condition and, 54, 66
Transponder
  to increase visibility, 19
  regulations concerning, 163
Trim control
  before-takeoff check of, 155
  cruise performance and, 122
Turbulence
  avoidance of, 38
  defensive steps during, 39
  and fatigue, 237
  intensity of, 37
  passenger comfort and, 36
  and selecting a route, 82-88
Turns about a point, 82-88

Unavoidable instrument conditions
  and flying over the top, 32-33
  and precipitation, 33

Unicom radio station, 215

Vanity, 54
Vertical speed indicator, 153
VFR over the top
  hazards of, 32
  precautions to take when, 30
Violation of regulations, 66

Weather briefing
  requesting a briefing, 152
  and selecting an altitude, 182-87
  and selecting a route, 187
  weather charts, 178
Weight and balance
  reading the charts, 69
  rule of thumb, 126
  and stall recovery, 125